SUSTAIN-ABILITY

BY

DANIEL J. MYERS

Cover photo: Aerial photograph of the Solar Electric Generating System (SEGS) in Kramer Junction, California.

Back cover: A Chinese woman heating water on her solar stove.

CONTENTS

DEDICATION

This book is dedicated to the enlightened men of science and good will who, over the centuries, have devoted their minds and energies to improve conditions that lift our lives above the ordinary.

"All eyes are opened, or opening, to the rights of man. The general spread of the light of science has already laid open to every view the palpable truth, that the mass of mankind has not been born with saddles on their backs, nor a favored few booted and spurred, ready to ride them legitimately, by the grace of God. These are grounds of hope for others. For ourselves, let the annual return of this day (July 4) forever refresh our recollections of these rights, and an undiminished devotion to them."

~ Thomas Jefferson, as quoted from his last letter before his death on July 4, 1826.

Soaring food prices. Food riots worldwide. Will millions of people starve?

What can be done about it? Let me tell you.

Gasoline prices are unstable. What can be done?

Read how an electric car that runs on lithium-ion rechargeable batteries can be plugged into an outlet in your garage. Each recharge is good for 250 miles and it costs only pennies a mile to drive!

Want to protect the environment and tap into an unlimited source of free energy? Learn about geothermal energy.

Want to cook out in the back yard using a solar stove? If so, a new sun powered parabolic stove may be just what you will need. This book tells you all about it.

Southern California is suffering from the worst drought in decades. What can be done?

Learn how fresh distilled water can be produced by boiling ocean water using the power of the sun.

When ocean water is boiled, billions of gallons of fresh water will be produced.

Deserts will bloom.

The wealth of nations will be measured by barrels of water flowing in their pipelines instead of oil.

Do yourself a favor. Read this book.

It is your one indispensable read.

CHAPTER 1: THE NEED

We have come a long way toward achieving optimal living conditions in many of the more technically and scientifically advanced countries of the world. A long life with good food, water, medicine, and a safe environment is the norm for most who live in the so-called "first world."

Nevertheless, there is extreme poverty and economic slavery with deplorable living conditions in most parts of the world, particularly in the under developed nations, the so-called "third world."

If we are to overcome the sad legacy of the poverty plaguing most humans on earth, we must ask ourselves whether or not we want, or even care, to help those who are in such drastic need of our help.

The greatest need is to help the downtrodden become self-sufficient in the production of food, safe drinking water, and enough electricity to light their homes and run their industries. Food, water, and energy are the components most needed and desired by those stricken with a pervading poverty of life and spirit.

We need to put pressure on the repressive regimes of the third world to allow their citizens the rights and freedom of association and the right to accumulate and keep wealth in open markets.

The task ahead will be much easier if free and open markets are allowed to drive up third world standards of living. In order to succeed, the enterprise must be profitable to both the givers and takers.

PROBLEMS OF THE THIRD WORLD

The problems of the third world (Less Developed Countries: LDCs) are many and will not go away. They are greed, avarice, racism, tribalism, ignorance, religious fanaticism, and radical or incompetent economic systems and policies.

The hope of third world people for a better life has caused the largest mass migration in the history of the human race. Third world emigrants are invading developed nations in search of a better life for themselves and their families.

Of course, the best solution would be to raise the standard of living in third world countries so that people could stay at home; however, the repressive regimes that control the economy of the nations where these unfortunate people live will not let them have a share of the wealth. The outlook for improvement is dim, indeed.

The global recession of 2008—that the third world is now suffering through—was made worse by real estate speculation in the United States. The debacle was caused by Representative Barney Frank (Massachusetts Democrat) and Senator Chris Dodd (Connecticut Democrat) who pushed Fannie Mae (Federal National Mortgage Association aka FNMA) and Freddie Mac (Federal Home Loan Mortgage Association aka FHLMA) to give mortgages to home buyers who had no ability to ever make good on the monthly payments. Both Fannie Mae and Freddie Mac were public companies traded on the New York Stock Exchange (NYSE). Fannie Mae and Freddie

Mac went bankrupt. The U.S. Government took over. Stock holders lost all their money. Chaos on the New York Stock Exchange caused by these worthless or subprime mortgages sent the NYSE on a downward spiral of over 40 percent between the fourth quarter of 2007 and the 4th quarter of 2008: from 14,000 to 8,000.

The market speculation was hyped up because of the easy money that glutted the marketplace. During the years 2001 through 2004, Federal Reserve Chairman Alan Greenspan flooded the U.S. with cheap money by reducing the discount rate as low as an astounding (1) percent. Greenspan testified before Congress on October 23, 2008 that the meltdown of the economy "revealed a flaw in a lifetime of economic thinking that left him in a state of shocked disbelief." He called the banking and housing chaos a "once-in-a-century tsunami." [1]

CONFLICT OF CIVILIZATIONS

Like an invading army, the third world (Less Developed Countries or LDCs) people invading Europe are mostly uneducated Muslims from the Mideast and Africa who bring with themselves a way of life different from that of Western civilization of the first world (More Developed Countries or MDCs). To further complicate matters, they refuse to adopt the culture and language of the invaded nations. In Europe the invaders bring a new religion—Islam—that confronts and competes with Christianity of the West. Consequently, Muslim ghettos that have kept their old ways, laws, language, and religion have sprung up in Europe, particularly in France and Germany. The

Muslims refuse to integrate into the culture of their hosts, thus causing a cultural malaise that is beginning to spread like a cancer within the body politic of both countries. If current trends do not change, it is possible that Europe will become Islamic by 2050, as the birthrate among Muslims surges and French and German population stagnates and ages. In Russia, for example, the indigenous population is already in decline. [2]

The Latin American invasion of the United States runs parallel to that of the Middle Easterners and North Africans who have invaded Europe, except that there is no religious conflict of interest.

It must be noted in passing that Asian nations do not have the problem of being overrun by their neighbors. For example, the Chinese and Japanese do not allow it and we can be sure they never will.

THE SECOND WORLD

By adopting open market economies and free international trade policies, China and India, the so-called "second world," have significantly improved the standard of living of their citizens. Remarkable economic advances have taken place within the past few decades. Let us hope that the enlightened economic policies of the first and second most populated nations on earth—China and India—will be an example to show third world nations the proper path to follow in order to achieve similar economic results.

TRANSPORTATION AND COMMUNICATION

I have not dealt with the need to develop transportation and communication systems within the third world because these services are already highly developed and cheap. They would become available in abundant supply when the basic need for food, water, and energy is satisfied. In fact, the ubiquitous cell phone seems to have taken over as the communication instrument of choice.

An electric car invented by Tesla Motors of San Carlos, California reaches a speed 135 miles per hour in about four seconds. It is driven by lithium ion batteries and has a range of 250 miles. The car is environmentally friendly, recharges on house current and operates at about one or two (!) cents per mile. The implication, of course, is that it won't be long before gas guzzlers will be a thing of the past. Some day in the not-too-distant future, this car will be the commuter vehicle of choice.

CHAPTER 2: FOOD

Our need for food is absolute.

FARMING GIVES BIRTH TO CIVILIZATION

The first need for human survival is an adequate and reliable supply of food. Without it we starve.

Without enough food, our existence becomes a life-or-death struggle to eat. People who cannot find enough to eat die. If crops fail or the supply of food is cut off for any reason, the death rate increases until enough people have died that there is enough food to go around for those who remain. This basic law of supply and demand for food has been with us from the beginning. It will stay with us to the end.

For a city to survive, farmers living in the countryside must grow more food than they can eat and be willing to sell what is left over to the people living in the city. City folk will starve or have to get out of town if the farmers cannot or will not grow a surplus of food that can be sold to them in the local marketplace. All of us who live in a city should think carefully about where the next meal would come from if the food stores and food markets could not sell us our daily bread.

Farming started about ten thousand years ago when farmers were able to grow more food than they could eat. This is possible because a single grain of some plants, such as wheat and rice, produced ten, twenty, or more new grains of wheat or rice at harvest. It was soon discovered that planting a crop in the same field year after year would yield a surplus year after year. As the producer of excess food, the farmer became a valuable member of his community. The land on which he grew his crops became a valuable asset.

The first known place where farming took hold on a regional basis was along the Tigris and Euphrates Rivers in southwest Asia. Other areas where farmers were able to produce excess food on a regional basis were along the Nile River in north east Africa and the Yellow and Yangtze Rivers in eastern China.

The development of settled (plow) farming is the most important event in human history. It ended the daily struggle to find food to eat. The days of the hunter and gatherer were over. Slash-and-burn farming dwindled as a way to grow food. This momentous event marked the beginning of civilization and recorded history. It lifted the veil of ignorance and bestiality that was our millennia of unrecorded history. Up until that time man was a brute: fighting, taking, and killing to get food for himself, his family, and tribe. The sowing and reaping of crops that yielded more food than the farmer could eat changed the course of human history. It made possible the development of cities.

FOOD SURPLUS AND POPULATION GROWTH

Plentiful food is a precious gift. It let people who were not farmers come together in communities, some of which eventually grew into big cities. In turn, the excess of food increased the birthrate, and as the result of rapid population growth, the inevitable shortage of food followed.

A food shortage can take place if there is a crop failure, usually the result of bad weather. War and pestilence are other well-known causes of food being in short supply. No matter the cause, a food shortage always leads to famine and death until the balance is restored between the shortage of food and the abundance of people. Early recorded history is characterized by a rhythm of checks and balances imposed upon population growth through the misery of feast and famine. The slow growth of agricultural surplus mandated a similarly slow growth of people. It is a remarkable fact that it took more than 10,000 years from the beginning of settled (plow) farming for the population of the world to reach one billion souls about 200 years ago. Yet in the reasonably short period of time since then that number has increased more than six times to over six billion people.

The Population Bulletin of September 2008 gives us the following World Population Milestones:

1,000,000,000 in 1804

2,000,000,000 in 1927 (123 years later)

3,000,000,000 in 1959 (32 years later)

4,000,000,000 in 1974 (15 years later)

5,000,000,000 in 1987 (13 years later)

6,000,000,000 in 1999 (12 years later)

This table highlights the rapid population growth since 1959. The numbers are astounding. The population of the world doubled from three to six billion people between

1959 and 1999, a time lapse of only forty years. The next milestone—7,000,000,000—is expected to be passed in 2011 or 2012.[3]

To feed this explosive quantity of more than three billion people since 1959 means that there was an equally explosive increase in the surplus food supply. Implicit in the figures, but not brought out in the chart, is the fact that about 10,000 years elapsed from the beginning of settled (plow) farming until there was enough surplus food to support a world population of one billion inhabitants.

WILL WE RUN OUT OF FOOD?

If the rate of population increase continues at the 12.6 percent rate that it grew in the 1990s, the population will double to about 12,000,000,000 by 2050. The question that comes immediately to mind is, "Can we grow enough food to feed the world's burgeoning population?" The answer to this question is not a simple "yes" or "no," because many factors will enter into the equation that gives us the answer at some future date. The answer depends upon whether or not we continue to make the genetic changes that produced the "super" strains of high yield disease-resistant grains that took place during the last half of the twentieth century. It was this so-called "green revolution" that gave us the food that permitted such amazingly rapid population growth in such a short period of time.

At one point scientists had believed—and the United States government agreed—that genetically engineered or genetically modified (GM) food was healthy for human consumption. Still, resistance to the growing and marketing of genetically engineered food increased, and even has support in some European countries as an expression of national policy. In the U.S. there is a growing preference for so-called "organic" food. Nevertheless, hungry people thrive on genetically engineered food without suffering any known ill effects.

RISE OF THE ALARMISTS

In addition to the food "purists" who insist upon eating no genetically engineered food, there are the "alarmists" who say that there are too many people on earth. The alarmists have broadcast a prediction of catastrophic collapse of the human race for centuries. There are also a growing number of "earth first" adherents who want to destroy most or all people because, they claim, that humans befoul the pure earthly environment.

Thomas R. Malthus (1766-1834) an English economist published his essay on population in 1826. In it he put forward the view that population, when unchecked, increases at a geometrical ratio while subsistence increases only in an arithmetical ratio. Malthus asserted as a fact that population always increases up to the limits of the means of subsistence. Population is prevented from increasing beyond these limits by the positive checks of war, famine, and pestilence and by the influence of misery and vice. [4]

Malthus can be said to be the first of a long line of alarmists who still write and sell volumes about the doomed fate of mankind as a result of the over population of the planet. Fortunately, food surpluses have forestalled the dire predictions of the alarmists, especially during the green revolution of the last half of the twentieth century.

About one hundred years before the birth of Thomas Malthus the population of the world passed the half billion mark. The population increased to one billion after another 150 years. This doubling of world population to one billion by 1804 was the beginning of the modern revolution in farming. Ten thousand years went by from the time that man first put plow to earth until the output of his labor could support a population of one billion people. Many factors contributed to the slow development of agriculture in ancient times. Among these were: absentee ownership of the land, restrictive use of the food surplus, primitive tools, and the labor intensive nature of farming.

ANGLO-SAXONS, THE STEEL PLOW FARMERS

The history of modern farming is an Anglo-Saxon story. When the Roman Empire collapsed in the fifth century AD, their legions withdrew from Britain, leaving the islands open to all comers. Among the first settlers were the Angles and Saxons. They were tribal farmers and fishermen living around the mouth of the Rhine River. They loaded up their boats with their families and steel plows and sailed across the channel to resume farming and fishing in Britain. With the Romans gone, most of the land was open and not farmed. The Anglo-Saxons proceeded to conquer Britain with their plows. They built family homesteads, and most important, they established laws under which land ownership was the property of the farmer who farmed it. Their system of agriculture was so successful that it expanded throughout the British Islands.

So successful, in fact, that Britain became known as “Angle-Land ”or as we say today, England.

The Anglo-Saxon method of farming changed drastically in 1066 with the arrival of William the Conqueror. William took possession of all the land of England and parceled out large tracts of it to his lieutenants who were entrusted to manage them in his name. This system of land management was a throwback to the method used in France where William grew up. These large land holdings are called manors. The manorial system of management gives ownership of the land to one man (the lord) who employs the farmers as his serfs. Vestiges of the manorial system still survive in Europe.

Six hundred years after William the Conqueror subdued them in England, and a thousand years after taming Britain, the Anglo-Saxons set sail again with their families and plows to tame America. This time, however, they secured their freedom and posterity by kicking the King of England out of America. To replace the monarchy, they established a representative form of government that guaranteed a broad range of legal protections, human rights, and the freedom to engage in open commerce. Important among these rights was the right of farmers to own land and be protected legally in its ownership.

The abundant availability of good farmland and the right to own and build homesteads on this new land was the incentive that inexorably drove the Anglo-Saxon farmers westward. They broke the sod with their steel plows as they went until they came to the shores of the Pacific Ocean,

3,000 miles from where they had started. The movement was rapid. The entire expanse of land that is now the United States was under the plow and actively farmed by the middle of the nineteenth century.

The Anglo-Saxon homestead in the America of the eighteenth and nineteenth centuries was an efficient marvel of self-sufficiency for its day. Families cleared the land with their axes and saws, using the trees to build homes (usually a log cabin if trees were available) and barns for the horses and cows.

Crops were planted during the spring, harvested in the fall, and stored in root cellars for eating until the next planting. In the autumn animals, usually pigs or steer, were slaughtered to provide an annual supply of meat for the family. Hunting provided wild game and birds for the dinner table.

The women took care of household chores, prepared meals, tended the vegetable garden, canned food for the winter, made soap and milked the cows. In addition, they used their spinning wheels to make clothes for the family.

The men plowed the fields, planted and harvested crops, built fences and roads, cleared new ground for planting, and constructed new buildings as needed. The farmer was a blacksmith who used his billows and anvil to make such things as iron tires for wagon wheels, hinges, hoes and many other implements for family use.

Figure 1-1. The Ratcliffe (my mother's family) Homestead in Little Valley, Tazewell County, Virginia on an October morn in 1928. Note the rail fences and corn shocks. From left to right note the beehives, Model T Ford, and family members, smoke house, blacksmith building over the root cellar, outhouse, barn, corn crib with wagon and the log cabin built about 1885. Photograph by Daniel J. Myers, Sr.

THE INDUSTRIAL REVOLUTION

Anglo-Saxon farmers who came to America were beneficiaries of the industrial revolution pioneered by other Anglo-Saxons who never left England. The industrial revolution in England coincided with the expansion of farming in America. It was characterized by new methods of producing iron in quantity, construction of textile mills that made large quantities of cloth available at low cost, invention of the steam engine, the building of railroads, and the construction of ships made of iron. These are but a few of the marvels that came out of the industrial revolution.

Perhaps the most important aspect of the industrial revolution was the fact that its beneficiaries welcomed it and were anxious to use the new inventions to improve their lives.

As beneficiaries, the American farmers soon had new mechanical threshers, plows, and combines pulled along by steam engines. New railroads crisscrossed the countryside to haul food to the many growing cities across the country. New machinery flowed from the factories in the new cities to the waiting farmers. Simply put, the Anglo-Saxon farmers of America mechanized farming.

The efficiency realized in food production created a surplus of food that exists to this day. That is to say, the increase in demand for food has not yet caught up to the rate at which the farmers are able to grow a surplus. Keep in mind that the world population is at six billion and growing.

A new plateau in farm mechanization was attained after the turn of the twentieth century when the internal combustion engine (tractor) replaced the heavy, lumbering steam engine as a source of power. New and more efficient farm machinery rolled out of factory doors onto the farms of America. There was new machinery such as plows, planters, mowers, binders, balers, cotton gins, cotton pickers, corn pickers, threshers, combines, tillers, harrows, discs, spreaders, sprayers, and more, as well as specialized machinery designed for specialty producers such as orchard owners, tree growers, cattle ranchers, sheep herders, and poultry farmers. Along with the new machinery that flooded onto American farms, low-cost manufactured goods such as shoes, dresses, suits, cloth, furniture, and sewing machines came in from the cities.

EDUCATING THE FARMERS

In the last half of the nineteenth century, the United States established Land Grant Colleges by requiring each of the states to build a college or university. According to the Congress that passed the first act creating the Land Grant Colleges in 1862 (Abraham Lincoln Administration), the purpose of these institutions was to educate the "industrial classes" in "branches of learning as are related to agriculture and the mechanical arts."

The Land Grant Colleges program has produced some of the best universities in the world. Among these are such outstanding universities as Ohio State, Iowa State, Pennsylvania State, Virginia Tech, Texas A&M, University of Georgia, Cornell in New York, the University of California System, University of Illinois, Michigan State University, and the University of Wisconsin. In fact, each state of the U.S. has its own Land Grant College. New acts of Congress still fund and define the program.

An important feature of the Land Grant Colleges Act was the creation of agriculture research stations and the granting of degrees in agriculture. The results of this research and the knowledge of the degree holder was taken directly to the farmer by a nationwide system of county agents.

Because there are thousands of counties (3,381) across the country, the County Agent is the farmer's neighbor and expert advisor who gives useful and free information on soil, drainage, fertilizer, crop rotation, irrigation, planting, tillage, harvesting, marketing, educational opportunity, and so forth. A Land Grant College studies and researches new and better ways to increase crop yields. The County Agent puts that new knowledge at the fingertips of the farmer.

The direct application of the latest research to improve agriculture production combined with the mechanization of farming and factory output of a cornucopia of low-cost goods put an end to the self-sufficient farm homestead. It became cheaper to buy manufactured goods than make them at home. The demise of the independent farm family marked the end of an era. First, the family was no longer able to survive on its own, but was dependent upon on the goods produced in the cities. Second, the mechanization of the farm reduced the need for large families to work the land. This resulted in a dramatic drop in the size of farm families and forced the members of some families to move into the cities to find jobs.

THE RESEARCHERS

In 1875 horticulturist Luther Burbank (1849-1926) moved from Massachusetts to Santa Rosa, California. He built a greenhouse and began to breed numerous varieties of fruit, vegetable, and flower plants, many of which are still grown. Burbank performed hybridization or selective experiments with thousands of plants representing nearly 200 genera. Burbank's work, which involved the raising of hundreds of thousands of individual plants, was not conducted to prove or test scientific theories or to make scientific discoveries, but had as its sole aim the production of more and better varieties of cultivated plants. In addition to these tangible attainments, however, his efforts gave great impetus to the development of plant breeding. From the viewpoint of science, this stimulus was perhaps the most important result of his work. [5]

Burbank is the pioneer plant experimenter who gave us many new and better plants and showed the way for scientists of the twentieth century to go beyond his method of selective breeding. When he died in 1926, his brilliant plant breeding techniques continued to be used by horticulturists to breed new strains of plants, particularly grains, with increased yield that were immune or resistant to many of the common plant diseases.

In 1953 James D. Watson, an American scientist, and Maurice H. F. Wilkins and Francis H. C. Crick, two English scientists, were able to determine the structure of deoxyribonucleic acid (DNA). DNA is a nucleic acid that is bound in double helical chains by hydrogen bonds between the bases, forming the basic material in the chromosomes of the cell nucleus. It contains the genetic code and transmits the hereditary pattern for all living species. [6]

In 1973 American biochemists, Stanley N. Cohen and Herbert W. Boyer, created the first genetically engineered organism. They did this by splicing together DNA fragments from several different species of an organism, mixing the traits of each to create a previously unknown variation. According to the National Inventors Hall of Fame located in Akron, Ohio, this accomplishment, known as "Recombinant DNA Technology," is the basis for modern biotechnology. It is better known to the world as "gene splicing" or "genetic engineering."

TOO MUCH FOOD AND TOO MANY PEOPLE

By 1960, the United States was ready to give the world a gift of immense importance: the green revolution. Because of the hearty strains of food plants produced by selective breeding and gene splicing, world food production increased rapidly and continues to grow in the twenty-first century. Population increase, though rapid, has been unable to catch up with a huge surplus of food. It is noteworthy that this gift was given willingly and at no cost to anyone.

THIRD WORLD RAPID POPULATION GROWTH

In mid-2008, world population stood at 6,700,000,000, up from 6,000,000,000 in 1999. During the twentieth century, nearly 90 percent of the growth took place in third world (LDCs) countries, classified as such by the United Nations—all countries in Africa, Asia (except Japan), Latin America and the Caribbean, and Oceania (except Australia and New Zealand). This remarkable growth resulted from an unprecedented decline in death rates in the third world brought about by the spread of public health measures, health care, and disease prevention, particularly after the end of World War II in 1945.

These improvements evolved over centuries in the first world or more developed countries (MDCs), but the third world countries were able to benefit from them virtually overnight.

The imbalance in population growth seen over the last century will only intensify in the years to come. Between 2008 and 2050 virtually all population growth will take place in the third world. Overall, the small amount of population increase projected for the first world will take place in the U.S. and Canada, but most of that growth will probably be due to immigration from the third world.

While the LDCs are projected to increase from 5,500,000,000 in 2008 to 8,100,000,000 in 2050, the MDCs are projected to grow from 1,200,000,000 to just 1,300,000,000. [7]

FIRST WORLD RESPONSE TO THIRD WORLD INVASION

Programs to protect the ethnic, religious, and cultural values of Western civilization were set up by European countries. As of 2008, the United States had an estimated 11,000,000 to 12,000,000 illegal immigrants from Latin America, mostly from Mexico. Illegal immigration had become a politically charged issue in the United States. Some politicians wanted to give the illegal immigrants U.S. citizenship, but a populist uprising nixed the effort and demanded that the U.S./Mexico border be protected.

POPULATION GROWTH AND FOOD GROWTH

Future population growth is certain. The certainty of future food surpluses is speculative, but not nearly so much as that of population growth. It is an absolute certainty, though, that population growth will be tied directly to and depend upon the production of surplus food supplies. As more food is grown there will be more corresponding mouths to feed. The development of hybrid and genetically modified plants offer the best hope for a continually increasing food surplus to feed the new billions that will be born.

From the perspective of today, there appears to be little or no cause for concern about the growth of world population because there is still a glut of surplus food due to the surge in food surpluses caused by the green revolution of the last half of the twentieth century.

FOOD AND POLITICS AND FAMINE

To be sure, many in the world still go hungry. There are two reasons why people starve to death. One is because of crop failure and the other is created by manmade famine.

There are several examples of famine related in the Bible. Each time their crops failed, the Jews went down to Egypt to get food. Century after century the Nile River supplied the ancient world with a reliable surplus of food that made Egypt one of the richest countries of the ancient world. Egypt, "the gift of the Nile," produced an abundance of food that supported a varied and advanced civilization. Ancient Egyptian civilization was a remarkable achievement.

In the 1930s there was massive drought in the United States, which persisted throughout much of the decade. Strong winds blew away the topsoil and created a desert in the southwestern United States.

It was called the "Dust Bowl." About 50 million acres were affected, especially in the state of Oklahoma. The drought forced thousands of farmers, the so-called "Okies," to abandon their land and migrate to California.

The big killer is manmade famine. Natural famines come along slowly and can usually be overcome, but one caused by man is quick and deadly. Manmade famine is created to satisfy political, racial, or religious goals. There are many historical examples of this.

One of the most notable examples of a politically engendered famine was that inflicted upon the Ukrainian people by the Union of Soviet Socialist Republics (USSR) Dictator Joseph Vissarionovich Stalin (ne Dzhugashvili) in 1932 and 1933. The Communist Stalin imposed collectivization or communization of Ukrainian farms. The kulaks (small farm owners) opposed the government seizure of their land. Their opposition enraged Stalin who engineered a famine by the confiscation of Ukrainian grain. The result was a famine that killed 6,000,000 to 7,000,000 people. [8]

A second notable example of politically motivated famine took place more recently in The People's Republic of China (PRC). After the defeat of Japan in 1945, war broke out between the Chinese Nationalists and Chinese Communists. The Communist Army under the leadership of Mao Tse-tung vanquished the Nationalist (Kuomintang) Army under Chiang Kai-shek and proclaimed the PRC on October 1, 1949. In 1958 Mao initiated the 'Great Leap Forward," the purpose of which was to bring together the large rural communes (farms) with a build-up of village industrialization. Village industrialization forced farmers to leave their farms and move to the cities to help in the industrialization process. As a result, food shortages created a disastrous famine in which an estimated 36,000,000 people or more starved to death. For the glory of socialism, Mao not only starved millions of Chinese, he executed tens of millions more and carried out genocidal warfare against the Tibetan people. [9]

AIDS, POLITICS, AND FAMINE IN AFRICA

Africans are destined to suffer most from future famine. Starvation is rampant in Africa now and millions will starve unless the United Nations continues to provide the increasing amount of food that will be needed to support its burgeoning population growth. The huge quantities of food that will be needed are available only from the West. In addition to a shortage of food, the situation is made worse by the AIDS epidemic that is devastating the people of black Africa south of the Sahara Desert.

Equally as devastating as famine and the AIDS epidemic is the continuing plague of political despots who rob the economies of these unfortunate nations and blame their failure on the West and Colonialism (white people).

Indeed, Colonialism is to blame. The European Colonialists developed the economic infrastructure of Africa over more than 200 years ago, beginning in the seventeenth century. Cities and harbors were built, railroads were laid, mines and factories were opened, universities were established, and land was cleared for farming. The thriving economies of these African colonies generated food surpluses which, naturally, led to a compensating increase in population.

The people prospered. But prosperity was not enough. With their new prosperity and its European educated elite, the Africans began to agitate for an end to the "exploitation" of their white rulers. By the middle of the twentieth century, the Europeans agreed to end Colonialism and began a process of turning their ruling authority over to African blacks. The process moved rather rapidly, taking only a few decades to complete. However, the black leaders were unable to keep their economies growing. The failing economies and the increasing population growth created food shortages that were made worse by incessant tribal warfare and pestilence in the form of a rampant AIDS epidemic.

In many respects the AIDS epidemic is a more serious economic problem for Africans than famine. Food can be imported from the surpluses available in the West to keep the people of Africa from starving, but there is no cure for AIDS. No one is optimistic that a cure for AIDS can be found in the near future.

Of the 33,000,000 AIDS infected people in the world, 22,000,000 are in sub-Sahara Africa according to a United Nations AIDS, 2008 Report on the Global AIDS Epidemic. The numbers show no sign of abating. The economies of black African nations are negatively affected by the AIDS pandemic because AIDS, a sexually transmitted disease, infects young men and women during the most economically productive years of their lives.

On July 30, 2008, President George W. Bush signed into law a plan that allotted $48,000,000,000 over the next five years for disease treatment and research. It built on the five-year, $15,000,000,000 President's Emergency Plan for AIDS Relief. It earned the President praise for acknowledging the AIDS epidemic and putting forth legislation to combat it. The AIDS epidemic claimed 2,000,000 deaths in 2007, most of them in sub-Sahara Africa. President Bush was condemned by Democrats for not including malaria and tuberculosis in his previous bills. The bill earned overwhelming bipartisan support in the U.S. House of Representatives and Senate. [10]

As bad as it may seem, the AIDS infected population of 33,000,000 as compared to the world population of 6,700,000,000 is a drop in the bucket at 0.0049 of the total world population or 0.49 percent. With such a small segment of the world population involved in the AIDS epidemic, the need for increasing food surpluses continues despite the ravages of the AIDS pandemic in sub-Sahara Africa.

Eventually the AIDS epidemic will subside as natural human immunity creates its own defenses against the disease. There is also the possibility that Western science will find a cure for the disease.

RACISM, TRIBALISM, AND RELIGIOUS INTOLERANCE

Unfortunately, there is no chance that there will be a cure for the racism and religious fanaticism that continues to disrupt the supply of food.

Two examples illustrate the point. In Sudan, the south is black Africans who are Christian and animist, whereas the north is Arab and Muslim. The Arabs sell the blacks into slavery. Additionally, the Arabs carry out government sanctioned military operations against the black Africans and have cut off their food supplies. An estimated 500,000 have died in war and famine.

Secondly, in Zimbabwe the black dictator, Robert Mugabe, authorized his family members and black political cronies to forcibly take over the land of white farmers. For many years Zimbabwe was an exporter of food, but now a shortage of food has occurred, the economy has broken down and hyper inflation has taken hold. At present, Zimbabwe is sustained with food from the West, without which, millions of black African Zimbabweans would starve.

COUNTER REVOLUTIONARIES

The green revolution that has allowed the population of the world to double since 1959 has also seen the birth of a virulent and violent counter-revolution. These activists oppose population growth, advocating reducing the number of humans in the world in order to "save the environment." They continually preach gloom and doom by predicting the imminent food shortages that will cause millions to starve.

Thomas Malthus started the alarmist ball rolling with the publication of his essays on population in 1826. Paul Ehrlich, a biologist at Stanford University, continued Malthus' doomsayer tradition with the publication of his book, *Population Bomb*, in 1968. The book warned the world of the immediate danger it faced. Ehrlich's logic was simple: A growing population increasingly consumes the earth's finite resources. Accordingly, this left humanity with only three options: (1) stop producing, (2) stop consuming, or (3) die from starvation.

His *Population Bomb* begins, "The battle to feed all of humanity is over… hundreds of millions of people are going to starve to death." In 1969, Ehrlich added, "By 1985 enough millions will have died to reduce the earth's population to some acceptable level like 1.5 billion people." The same year, he predicted in an article entitled "Eco-Catastrophe!" that by 1980 the United States would see its life expectancy drop to forty-two because of pesticides, and by 1999 its population would drop to 22.6 million. In the mid-seventies, with the release of his *The End of Affluence*, Ehrlich incorporated drama into his dire prophesies. He envisioned the President dissolving Congress "during the food riots of the 1980s, followed by the United States suffering a nuclear attack for its mass use of insecticides." That's right, Ehrlich thought the United States would get nuked in retaliation for killing bugs.

The three million copies of the *Population Bomb* that sold were influential in the radicalization of conservationist organizations such as the Sierra Club, and in creating the new ones like Greenpeace. Since his foray into environmental tomfoolery, Ehrlich's predictions have been consistently and tragically wrong for four decades and counting. [11]

Events of the past forty years have totally refuted Ehrlich's wild predictions. Logically, we would expect an end to such irresponsible predictions by our so-called educated elite. But this is not the case. Ehrlich still has a world wide following of activists eagerly predicting the imminent starvation of millions.

SOCIAL AND POPULATION ENGINEERING

Henry W. Kindall and David Pimentel wrote a monograph, "Constraints on the Expansion of the Global Food Supply" in May 1994. It was published by the Royal Swedish Academy of Science and carried on the Ehrlich tradition that predicted the inevitability of mass starvation. Kindall and Pimentel, a couple of U. S. academics, gave us a mid 1990s update on the food disaster that awaits the human race.

Of course, the world has a chance to avoid catastrophe if it will only adhere to the social, financial, and environmental disciplines prescribed by these two worthies. They make the logical statement, "It is not possible to make useful forecasts," but make them nevertheless.

Kindall and Pimentel predict wide spread starvation if we continue on our current course of food production, which they dub "business as usual" (BAU). The BAU scenario sees food production continuing to expand as it did in 1994 (and as it continues today). This scenario projects a world population of about 9.5 billion in 2050 by assuming that world population will increase at an ever-decreasing rate that adheres closely to the growth projected by the U.S. Census Bureau.

Under their BAU scenario, most developing nations suffer reductions in per capita grain production. In general, it appears that Africa, as well as China and India, will face particularly severe problems in expanding their food supplies in coming decades. The people of these regions will make up almost two thirds of the world's developing countries, and over half of the world's population in the interval between 2025 and 2050.

According to them, the United States appears to have the potential of generating food surpluses for some years, a potential it shares with parts of Europe to include Eastern Europe, Canada, and possibly other regions. But the authors note that even for the developed nations, "The longer-term prospects are unknown in view of difficulties which may appear later."

Kindall and Pimentel project a "pessimistic scenario" (PS) that adopts most of the assumptions of the BAU in which the world population rises to thirteen billion by 2050, but in which the developing countries (third world or LDCs) suffer decreases of as much as 10 percent in food production.

They write, "We believe that, in the period to 2050, the greenhouse effect and ozone loss could together depress grain yields on a global basis by 10 percent to 20 percent...While this represents nearly a 30 percent increase in grain production over 1991, it means per capita production would be down over 40 percent. There is, in this scenario, little hope of providing adequate food for the majority of humanity by the middle of later decades of the period we consider (2025-2050)."

Kindall and Pimentel also create an "optimistic scenario" (OS) in which the environmental, fiscal, and social wishes of the authors are carried out. In the optimistic scenario, the world population is stabilized at 7.8 billion in 2050 instead of the thirteen billion projected under the business-as-usual and pessimistic scenario.

They write, “Both the developed and developing nations would have to give high priority to food production and protecting the environment so as to maintain a productive agriculture. The developed countries would have to help finance these changes and also provide technology to the developing nations. At the same time, with diet shifts in the developed world, the 2050 population of 7.8 billion might be fed an adequate diet.” (Note: World population will pass 7,000,000,000 within the next year or two.)

Kindall and Pimentel write, “… the human race now appears to be getting close to the limits of global food productive capacity based on present technologies. A major expansion in food supply would require a highly organized global effort, (emphasized) by both the developed and the developing countries that have no historical precedent.

“As yet, a major commitment from the developed nations to support the needed changes is missing, and inadequate commitment in both developing and developed nations has been made for population stabilization.”

They conclude, “Governments so far appear to lack the discipline and vision needed to make major commitment of resources to increase food supplies, while at the same time reducing population growth and protecting land, water and biological resources. While a rough doubling of food production by 2050 is perhaps achievable in principle… the elements to accomplish it are not now in place or on the way. A large number of supportive policy initiatives and investments in research and infrastructure as well as socioeconomic and cultural changes would be required for it to become feasible. A major reordering of world priorities is thus a prerequisite for meeting the problems that we now face" (emphasized).

NEW ACADEMIC/SOCIALIST PARADIGM

That is it.

The Kindall and Pimentel study is the current academic/socialist elite statement of the problem with world food supply. Humans are "getting close to the limits of global food productive capacity." If we continue along the course we are on today (business as usual) the population will increase to probably 13 billion with "little hope of providing adequate food for the majority of humanity." On the other hand, if governments follow the authors' advice by implementing "a major reordering of world priorities:, the population will stabilize at 7.8 billion by or before 2050.

Kindall and Pimentel conclude that stabilization will not happen because "governments appear to lack the discipline." Please note that the authors see no solution to the impending starvation awaiting us by or before 2050 unless we deliver ourselves into the hands of bureaucrats, which they believe will not happen. If the authors believe, as they have written, that nothing will happen to save humanity from starvation, why, one would logically ask, did they bother to write it at all?

THE SOCIALIST SOLUTION

The answer, of course, is that they do expect something to happen to save us from starvation. The authors believe that the elitist blueprint they have laid out will inspire the academics, socialists, environmentalist and politicians to save the world from the certain starvation of millions of people. Their conclusion reinforces the "scientific" rationale of the radical socialists and environmentalists whose agenda is control and regulation of people's lives and economic activity.

MY CONCLUSIONS

The Kindall and Pimentel monograph is fifteen-years-old. But it is still the "bible" for activists. Food has become a lever used by socialists to justify their political and environmental extremism. The dire prediction of what disasters will befall the human race if their agenda is not followed is beginning to out-Ehrlich Ehrlich's doomsday predictions.

It is worthwhile to point out that the activists never present a scenario in which there is a positive solution to the problem of presumed future food shortages, such as how food supply can be increased as the world population grows. The solution is always negative. Their agenda calls for a worldwide political and social crusade to save the planet.

The agenda envisions reducing the human population, saving the environment, enforcing the Kyoto Protocol to reduce greenhouse gases, saving the ozone layer, saving endangered species, and forcing developed nations to give money, food and technology to under developed nations (primarily Africa).

The Kyoto Protocol excluded China and India, two of the most polluting nations on earth. The Kyoto conferees express a vague hope that China and India would follow the developed nations, and especially the United States, in adopting the Protocol sometime in the future. President William J. Clinton understood the mechanics of the scheme and rejected it by refusing to submit the Kyoto Treaty to the United States Senate for approval.

KYOTO PROTOCOL AS RELIGION

The Kyoto Protocol has assumed the mystic of a religion, so much so that the politically correct consensus will not allow anyone to disagree with it. Former Vice President Albert A. Gore, Jr. is the chief guru of the religion of global warming. He proclaims that green house gases will cause the temperature of the earth to heat up a few degrees within the next hundred years.

EXTREMISTS INVADE JOHANNESBURG

The political activism needed to set up a world-wide agenda that can put pressure on the wealthy nations of the world continues unabated. The World Wildlife Fund (WWF) issued its "Living Planet Report 2002" in July 2002. The report promoted the United Nations "World Summit on Sustainable Development" (WSSD) which was held in Johannesburg, South Africa from August 26 to September 3, 2002.

According to the WWF Report, "humans are currently

running a huge deficit with the earth—using over 20 percent more natural resources each year than can be generated—and this figure is growing each year." The report predicted that "based on likely scenarios of population growth, economic development, and technological change, that by 2050, humans will consume between 180 percent and 220 percent of the earth's biological capacity."

The WWF Report continued, "this means that unless governments take urgent action, by 2030, human welfare—as measured by average life expectancy, educational level, and world economic product—will go into decline."

Of course, there was no basis in fact for such wild speculation.

After making this dire prediction, Jonathan Loh, author of the report wrote, "We do not know exactly what the result will be of running this massive overdraft with the earth. What is clear, though, is that it would be better to control our own destiny, rather than leave it up to chance."

Mr. Loh continued, "At the World Summit on Sustainable Development (WSSD), world leaders will have a magnificent opportunity to address the root causes of our obvious failure to achieve sustainable development and set us on the path to a truly sustainable future."

The WSSD turned into a political spectacle of finger pointing and recriminations because then Secretary of State Colin Powell attended the conference instead of the sitting president, George W. Bush.

President Bush was urged not to attend the summit by more than thirty conservative activists who wrote in a letter to him, "Your presence would only help to publicize and make credible their various anti-freedom, anti-people, anti-globalization and anti-Western agendas. We also strongly support your opposition to signing new international environmental treaties or creating new international environmental organizations at the Johannesburg summit."

Fred L. Smith, Jr., president of the Competitive Enterprise Institute, said in an interview with the Associated Press, "Why would you want to go to a party when they want to throw pies at you? The fortunate thing is when 40,000 goofies get together, not much happens."

On the other hand, United Nations Under Secretary-General Nitin Sesai, who chaired the summit, told reporters, "There is a real sense of urgency. In many cases we are talking about slipping back. There is no major global event planned beyond Johannesburg which will allow us to retrieve lost ground. This is it."

CRITICAL REPORTS

The fact that President Bush did not attend WSSD upset environmental activists, particularly in view of the fact that he rejected the Kyoto Protocols on climate control.

"People around the world are seriously concerned that the Bush Administration is undermining the World Summit instead of working with other countries to benefit everyone," Sierra Club director Michael Dorsey said.

According to a Capitol Research Center report the

green groups were very unhappy with the outcome of the summit. Friends of the Earth (FOI) International Chairman Ricardo Navarro said that Secretary Colin Powell was "shameless and inadequate" and that the summit was "hijack[ed] by free market ideology." FOI director Charles Secrett ranked the Summit "the biggest political sell-out in decades."

The Sierra Club's Michael Dorsey complained that the summit hadn't been hard enough on private companies. When it let corporations in on the deliberations, summit organizers "did enshrine, for once and all, the role of corporations in the sustainable process."

"The Earth Summit has failed… to make corporations accountable for the devastation they bring," said Paul Horsman of Greenpeace.

A representative of the British aid group OXFAM called it "a triumph of greed and self-interest."

A spokesman for the World Wildlife Federation said the summit acronym WSSD should really stand for "World Summit of Shameful Deals."

UNITED STATES COUNTER ATTACK

All the hand-wringing was caused by the fact that the U. S. State Department vigorously argued the Bush case for more global development, and it blunted the public relations offensive of the environmental movement.

At the WSSD, U. S. officials staged a head-on rebuttal of the anti-American, anti-technology attack. Consider genetically modified (GM) foods. The U.S. believes genetic modified (GM) crops will help countries in Africa overcome their chronic food shortages. Even in a good year, an average African farmer, relying on primitive technology, can hope for a crop yield of 1.7 tons of corn per hectare (2.47 acres) planted. In the U. S., by contrast, biotechnology helps farmers get eight tons per hectare. If Africans had GM corn seeds that are immune to disease and impervious to insects, the continent might overcome starvation and malnutrition. Several African countries ban the cultivation of GM crops because European nations will not buy their crop surpluses. Groups like Greenpeace and Friends of the Earth support the ban.

Secretary of State Colin Powell's speech to summit delegates went straight to the point. "In the face of famine, several governments in southern Africa have prevented critical US food assistance from being distributed to the hungry by rejecting biotech corn which has been eaten safely around the world since 1995."

Andrew Natsios, Administrator of the U. S. Agency for International Development (AID), put the blame on Western green groups. He said, "The U.S. is not going to sit there and let these groups kill millions of poor people in southern Africa through their ideological campaign." That was the strongest denunciation that environmental extremists had heard in many, many years.

The strong U.S. response trounced its green critics in Johannesburg and offered a message of hope to everyone around the globe who wants to tap the power of new technology and the free market. At one point in the summit,

600 farmers from Africa and Asia marched through the streets of Johannesburg demanding the freedom to use biotechnology to grow better crops. Their future is brighter because of the Powell support and the Natsios condemnation of the activist critics.

THE WTO, THE EU, AND GM FOOD

More help is on the way. The World Trade Organization (WTO) on February 7, 2006, ruled in favor of United States biotechnology companies. The WTO ruled that European Union (EU) restrictions on genetically modified (GM) foods violated international trade rules. The ruling means that GM corn, soybeans, and cotton will be allowed access to the European markets. According to the *Wall Street Journal*, "The European Union is set to approve for cultivation a genetically modified potato used to make paper coating in 2007. However, dozens of other Biotech companies awaiting approval for dozens of other crops, worry that the potato decision might not represent a shift in European Union thinking."

Of course, the EU concern, now and in the past, is a political hoax designed to protect and subsidize European farmers, especially in France and Germany.

FREE MARKETS VERSUS SOCIALISTIC MARKETS

The ongoing worldwide political food fight is being carried on between free market adherents led by the United States and the former communists and socialists. As a form of government, Communism has lost all respectability; therefore, the communists and socialists have changed their agenda. They now call themselves environmentalists.

But their goal has not changed. They still want to control and regulate people's lives and economic activity. Their environmental message always bemoans the excessive number of humans who crowd the earth and they prescribe various ways that the nations of the world can get rid of them. Humans are portrayed as spoilers whose numbers, if not reduced through action by the governments of the world, will lead to environmental catastrophe and massive starvation.

However, these big, rich environmental organizations—such as the World Wildlife Fund—are not willing to sponsor a program as to how they would get rid of the excess of unwanted human beings that, according to them, plague the world's overburdened ecosystem.

To openly put forth a program mandating how the population will be reduced, and to specify who would be the first humans to be killed would destroy the prestige, credibility, and viability of such an environmental group.

RUSSIAN (SOVIET) KILLER ELITE

Of course, everyone knows that there is absolutely no way to carry out such a population reduction program without benefit of an all powerful overreaching world dictatorship with the power to kill people. We can be thankful that a worldwide killer elite has never existed.

A killer elite of Communists (the elite of socialism) did exist in the Union of Soviet Socialist Republics (USSR) from 1918 until 1989. The dictator of the USSR, Joseph

Stalin, killed Russians by the millions. No one knows exactly how many innocent souls were put to death. My belief is that there were tens of millions. Death was dealt out in labor camps, Soviet prisons, purges, and in-secret trials because life was cheap; furthermore, the Soviet killer elite considered it their duty to cleanse the USSR of undesirables. The carnage took an awful toll on the Russian people. The Russian nation comprises more land area within its borders than any other on earth, yet the population is less than 143,000,000 and declining. The Soviet census of 1939 listed the population as over 170 million people. [12]

In fact, the problem of a declining population has gotten so bad that Vladimir Putin announced a bonus, started in 2007, to be paid to women who give birth.

CHINESE COMMUNIST KILLER ELITE

Mao Tse-tung became Chairman of Communist China in 1945. The Chairman immediately set about to purge China of undesirables, especially the Nationalists that he had just defeated in a long military struggle. Millions died. Mao tried to fast forward the industrial development of China by forcing farmers to move into cities. A famine ensued and millions more died. Mao embarked on a war of genocide. Tibet and its people were the unfortunate victims of yet another Mao purge. Millions more died. Mao's "Great Proletarian Cultural Revolution" caused turmoil and confusion that convulsed the nation between 1965 and 1968. Millions more died. No one knows how many people were put to death by the Chinese Communists. I have seen estimates that 85 million were killed, but the total probably came to many millions more.

President Richard M. Nixon dramatically changed the United States relationship with China when he visited in 1972, the first president to do so. Since that visit, relations between the two countries has gradually improved, especially after the death of Mao in 1976. The wholesale slaughter of Chinese by the Communists is believed to have subsided since then.

CAMBODIAN KILLING FIELDS

Pol Pot (ne Saloth Sar), the General Secretary of the Khmer Rouge (Communist Party), was dictator of Cambodia from 1975 until 1979. During that time, he instituted an extreme version of agrarian Communism. People who lived in the cities were forced into the countryside to work on collective farms and in forced labor camps. The extremely harsh working conditions, malnutrition, poor medical care and executions, killed two million people—or about one third of the population of Cambodia.

ENVISIONING EXTREMIST POPULATION SOLUTIONS

In order to comply with the environmental extremists who insist on reducing the global population to their acceptable level between one and three billion humans, let us imagine how the slaughter of three to four billion people might be carried out.

Huge furnaces could consume millions of people to

generate electricity for the fortunate ones still waiting to be sacrificed.

Oil tankers that have been converted to haul people could drop hundreds of millions more into the ocean.

That old stand-by, the mass grave, could account for more millions.

I am sure the radical, fanatic, environmentalists can come up with many more ingenious ways to do the job, such as spreading deadly communicable diseases. Amazingly, some academics are already suggesting a program that would kill off millions by spreading a deadly communicable disease such as the ebola virus.

PROBLEMS WITH POPULATION REDUCTION AND CONTROL

We can get an insightful look at how environmental extremists think by examining an organization called Negative Population Growth (NPG) that has put forward a public program for population reduction in the United States. NPG is a national membership organization founded in 1972 to educate the American public and political leaders about the detrimental effects of overpopulation on our environmental resources and quality of life.

It advocates gradually reducing the United States population of 300 million down to 150-200 million. NPG claims that only with a much smaller population can we protect our fragile ecosystems, conserve our finite resources, and ensure that future generations will inherit a clean and healthy environment where all Americans can enjoy a quality standard of living. It advocates reducing the United States population by 50 percent by limiting immigration, taxing large families, paying cash bonuses to small families, shaping public opinion, public education, outreach programs, educating our political leaders, and supporting the efforts of grassroots activists. NPG believes that a world population of two to three billion would be optimal.

The NPG does not tell us how it wants to get rid of 50 percent to 60 percent of the people in this world.

ENOUGH FOOD FOR ALL?

For most people, food is one of the most boring topics of discussion in the world today. This is true because an abundance of food is still being generated by the green revolution and provides more than enough food for an expanding world population. Greater yields continue to flow from farms worldwide despite the fact that the United States and other nations have taken millions of acres of land out of agricultural production. This backlog of land, which can be brought into future farm production, means that there will be no starvation on the world scene any time in the near future, and that the population of the world will continue to expand because of the easy access to this food surplus.

The Congress has passed legislation mandating that part of our corn crop be processed to produce ethanol to be mixed with our gasoline supply before it can be pumped into gas tanks at filling stations. The Cable News Network reported on April 14, 2008, that the soaring cost of basic foods had caused food riots from Haiti to Bangladesh to Egypt and catapulted food to the forefront of the world's attention. Many people think the use of grain—such as corn—to produce ethanol is a shortsighted policy that the Congress should repeal.

Eventually, the increasing number of mouths to be fed will exhaust the food surplus unless the population stops growing. As noted above, Kindall and Pimentel envisioned in 1994 that under the best of conditions the world population, which was at 6,700,000 in 2008, would level off and stabilize at 7,800,000 by 2050. They predicted that world population would be 13,000,000 by 2050 if we continue "business as usual" and do nothing to stop the world's rapidly growing population.

Which will it be?

Will we have a stabilized world population or one that is growing too fast? The answer is not a "yes" or "no."

In the latitudes between 30-60 degrees north and south where almost all Caucasians and Asians live—except for India—many nations have already achieved zero population growth. In fact, the population in some countries is declining. The people who live in the earth's northern and southern latitudes produce prodigious surpluses of food, despite violent and unpredictable annual rainfall and temperature swings.

The famines that plague the modern world occur almost exclusively in the earth's "torrid zone," which is defined as the earth's surface encompassed by the area lying 30 degrees north and south of the equator. This is the surface of the earth occupied by the darker races of mankind and despite excellent soils, fertilizers, and mild temperatures starvation is a problem, especially in Africa. The population in the earth's "torrid zone" is growing out of control and there is no effective effort being made to bring it under control.

The current mantra demands that the rich Caucasians and Asians must give their food and wealth to the blacks. Of course, this silly idea does not solve the food shortage because free food from the rich nations enables the high birth rate of the poor nations to continue, creating a growing demand for more food which would, in turn, create a vicious cycle of larger and larger famines.

UNITED STATES FOOD CORPS

Is there a better way?

I think so. The United States needs to establish a Food Corps based along the lines of the Peace Corps. The Food Corps, however, would not be staffed by young immature volunteers but by mature, educated agricultural specialists. These agricultural professionals would be dispatched to host countries to train specialists in the latest scientific and ecologically safe methods of farm production. The mission would be to set up a self-sustaining infrastructure similar to the system of U.S. County Agents to assist farmers and supervise them in the best and most productive methods of farming. The goal would be to increase food production so that the host country can produce enough food to feed its people.

In addition to training agricultural specialists in host

countries where there are food shortages, the United States should make the establishment of an Agriculture University a condition for continued receipt of assistance from the Food Corps. These universities, fashioned after the Land Grant Colleges of the United States, would educate and train agricultural specialists and conduct research using the best available science.

The mission of these universities would be to set up and supervise a system of agricultural specialists in such numbers that they can work with farmers to help them increase food production. The goal would be to make that country self-sufficient in the production of food. The Food Corps would be established to operate on a flexible plan adapted to the needs of each host country.

FOOD PARKS AND FOOD TOWERS

As the population of the world increases, as it inevitably will, those nations that do not have shortages of food at this time should prepare for the day when shortages will occur if nothing is done to boost food production. This is especially true for nations that are essentially city-states. Here is a look at the approximate population in 2008 of the major metropolitan areas that dominated the better known city-states of the world.

Tokyo, Japan	35,000,000
Seoul, South Korea	21,500,000
Mexico City, Mexico	19,500,000
Cairo, Egypt	14,500,000

Manila, the Philippines	14,000,000
Buenos Aires, Argentina	13,500,000
Jakarta, Indonesia	13,500,000
London, England	12,000,000
Karachi, Pakistan	11,000,000
Istanbul, Turkey	10,500,000
Paris, France	10,000,000
Teheran, Iran	10,000,000
Dhaka, Bangladesh	9,500.000
Lima, Peru	7,500,000
Bangkok, Thailand	7,000,000
Bogota, Colombia	7,000,000

There are also big cities in the high population countries of Brazil, China, India, Russia, and the U.S.

I say that because of the inordinately heavy concentration of people living in the cities and nations listed above, some of them need to take "food insurance" against the future possibility of famine. The guarantee of an adequate and reliable food supply is especially important in those nations with little land to farm or infertile farmland.

Granted, there is little need at this time to worry about future food shortages, but the prudent nation should prepare now for the worst.

A national government would insure against famine by making and maintaining "food parks."

FOOD PARKS

Food parks would consist of plots of land given, sold, or leased to individuals (families). Terms of the ownership would specify that the individual must strive to maximize crop production by intensive farming using the latest scientific methods and best fertilizers available.

Each park would be supervised by an agriculture specialist. The specialist would be trained to help other farmers get the highest yields out of their crops.

In return for accepting the conditions of ownership, individual farmers would be able to sell their surplus on the open market.

Food parks could consist of a few acres or hundreds of acres in or around cities with a high density of population, or scattered throughout the countryside depending upon the need of each nation.

Properly managed, the food parks are a guarantee that each nation is assured a surplus of food to meet the uncertainty of future demand.

Climate and technology are important considerations when setting up food parks. In the warm climes, intensive farming could yield more than one crop a year, requiring the proper conditioning of the soil and use of fertilizer.

FOOD TOWERS

In colder climates, I envision that rich nations with advanced construction capabilities and technology would build "food towers."

Food towers could be a low-rise built to rotate so that all plants under cultivation would get equal exposure to the sun. The tower could be fitted with a roof that opens to the sky on sunny days.

The high rise tower would have solar panels or parabolic heat collectors (please see Figure 3.3) to supply interior warmth. Supplemental heating might be necessary during extremely cold spells in winter to ensure a year around crop yield.

It is easy to imagine that an entrepreneur could build a food tower in or near New York City that would be profitable in the market today. In fact, it is easier to contemplate that several food towers along the Boston to Washington corridor could be very profitable. Such a food tower—or towers—owner could supply nearby urbanites with the freshest, highest quality foods (GM and organic) available at a cost lower than lesser quality foods shipped in from Florida or California.

THE POSSIBILITIES ARE UNLIMITED

With the proper use of science, technology, and fertilizer to manage the means of ecologically safe production of food, nobody needs to go hungry; no matter how long it takes the population of the earth to stabilize, be it twenty-five or fifty or one hundred years or more.

What about overcrowding?

There is lots of living room on earth, the more the merrier.

People today are eating better, living longer, and

enjoying healthier lives than ever before.

I predict that life will get even better.

All we need to increase the world's farmland beyond our imagination, to make the deserts bloom, is water. To learn about how easy that would be, don't skip the next chapter.

CHAPTER 3: WATER

THE ELIXIR OF LIFE

Without water we die. All living things—plant and animal—must have water. Food and water are the two absolutely essential ingredients that are necessary to keep us alive. Whereas, we may live on the stored fat in our body without eating for weeks, without water we will die in a matter of days.

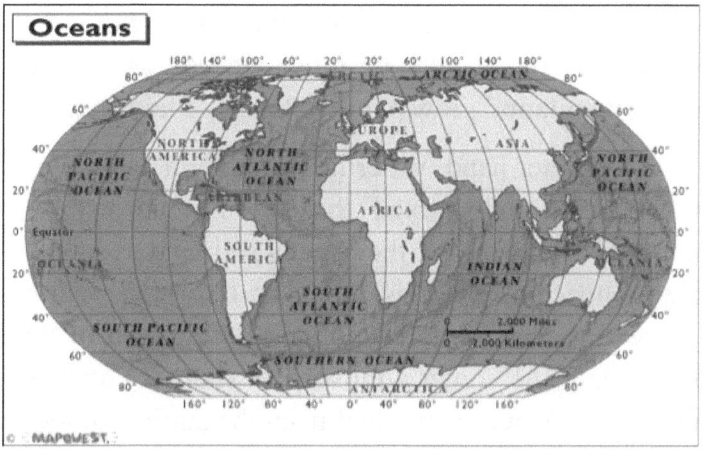

Figure 3.1 Oceans of the World Source: mapquest

About 71 percent of the earth's surface is covered by ocean water. The enormity of the volume of water in the oceans is seen from the Encyclopedia Britannica calculation that they contain 328,750,000 cubic miles of water. But humans cannot drink ocean water because of its high salinity (about 3.5 percent salt or sodium chloride—or NaCl written in its chemical format). It is deadly when substituted for fresh water. This deadliness is the source of the old sailor's lament, "Water! Water! Everywhere! Not a drop to drink!" Many have died of thirst upon the briny deep with nothing but water in sight from horizon to horizon. Fresh water that humans must have to live is found on the other 29 percent of the earth's surface where we live.

There is, however, a big difference between the availability of water and food. An adequate supply of food requires that farmers till the land and husband each crop to harvest. The process is capital and labor intensive and requires close attention to each crop, from planting, weeding, fertilization, harvesting to storage and preservation. The supply of food needed for the world's growing population grew over centuries of farming.

On the other hand, water is available in an amount that has been essentially fixed throughout recorded history. Water is distributed over the oceans and land as rain, snow, or ice dropped from clouds. Clouds are formed when water on the earth's surface is evaporated by the heat of the sun and rises into the cooler atmosphere where it condenses to form clouds. Water evaporated from the surface of the earth is fresh, drinkable water free of salt and all other minerals.

Water falls from clouds to earth when the saturation level of the water in the cloud is high enough to form rain, snow, or ice droplets that are heavy enough to fall.

WATER IS A SOLUTION FOR ALL PURPOSES

Water is an amazing substance. It is a molecule composed of two hydrogen (1H1) atoms and an oxygen (16O8) atom and is written in its chemical formula as H2O. It is familiar to us in all of the three forms that we find in nature; namely, as a solid (ice), liquid (water), and gas (steam). Water freezes at 0° Centigrade (C) or 32° Fahrenheit (F) and becomes a solid at all temperatures colder than this. Water is liquid between the temperature of 0°C and 100° C (212° F). At 100° C water boils and becomes steam or vapor above that temperature.

We must drink it to live. Water is the universal solvent.

We use water as a cleaning agent to dissolve solids, to wash clothes, to sterilize, to bathe; in fact, water is an inseparable part of our everyday lives. Water cools us when we are hot and warms us when we are cold. Water turns the turbines that generate hydroelectric power. Steam from boilers turns the turbines that generate electricity. Water is so versatile and is used in so many ways that we tend to take its utility for granted. Water is an amazing substance.

THE BIG PICTURE

The primary source of fresh water is evaporation caused by the direct rays of the sun striking the Pacific, Atlantic, and Indian Oceans along the earth's equator. The direct rays of the sun migrate between the Topic of Cancer (23° 26' N) on June 21st and the Tropic of Capricorn (23° 26' S) on December 22nd caused by the precession (wobble) of the earth on its annual orbit around the sun.

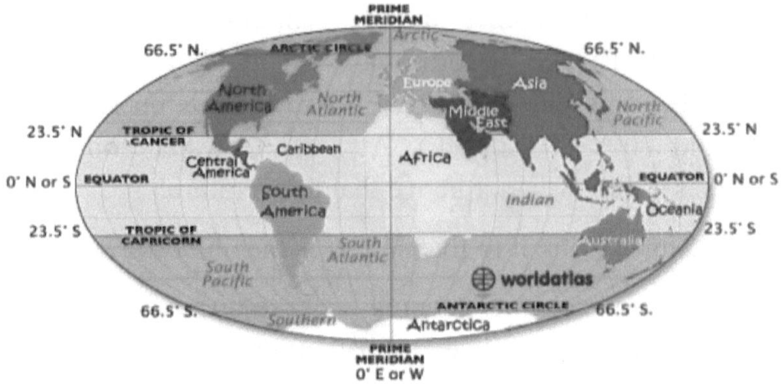

Figure 3.2 Tropics of Cancer and Capricorn
Source: world atlas.com

A gigantic, ocean-spanning column of hot water vapor boils off the surface of the ocean as it rotates under the direct rays of the sun. The water vapor rises above the ocean until it cools in the stratosphere (twenty to fifty kilometers in elevation).

Cooling increases the density of the water vapor, causing it to fall back to earth at 30° to 50° north and south of the equator. The latitude where the water falls back to earth depends upon the seasonal precession of the earth in its orbit around the sun. Winds distribute this water vapor throughout the earth's atmosphere where some of it eventually condenses into clouds and falls back to earth as rain, snow, or ice.

As this gigantic latitudinal column of hot evaporated water rises above the ocean it creates lower pressure along the northern and southern edges of the column at the surface of the ocean. High pressure is created by the cooler and heavier air north and south of the equator as it falls back to earth. The air from the areas of higher pressure (cold air) north and south of the equator flow into the area of lower pressure (warm air) along the northern and southern base of the ocean-spanning column of evaporated water.

This steady flow of air is called the prevailing or trade winds. The flow of these desiccating, or drying, trade winds as they rush back to their origin rob some parts of the earth's surface of its moisture, thereby creating deserts in both the northern and southern hemispheres.

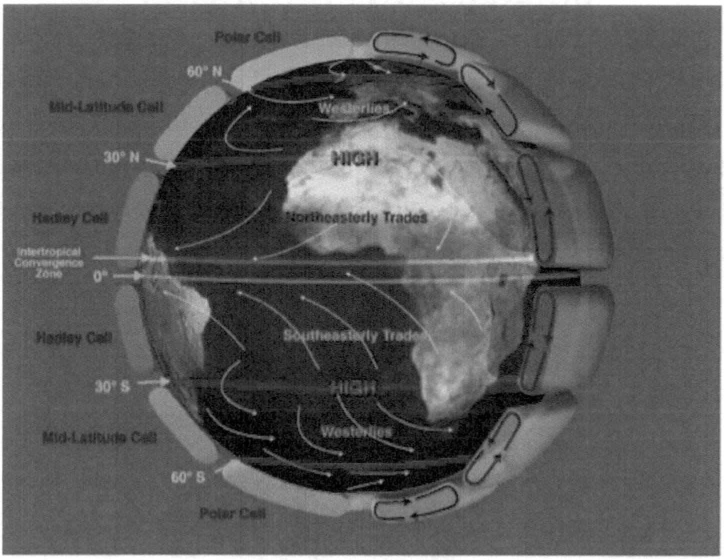

Figure 3.3 Trade Winds of Earth. Source: Wikipedia.com

The rotation of the earth causes trade winds to blow persistently from the northeast within the northern hemisphere and from the southeast within the southern hemisphere. As the trade winds reach the warm low pressure at ocean level, they are heated up and once again carry evaporated water into the stratosphere to repeat the cycle.

The distance between the north and south edges of this long, gigantic column of water vapor is called the "doldrums" because there is no flow of air. It is here that early sailing ships were stranded until it was discovered that it was easy to row across the doldrums to the trade winds on the other side.

SOURCES OF FRESH WATER

Some places on earth are natural storage sites for fresh water. These places create conditions favorable for human habitation. Lakes are the most obvious and familiar sites for the storage of fresh water. A lake is formed when water—either from rain, a stream, or both—fills up a depression in the surface of the earth.

Figure 3.4 The Great Lakes Source: teach. GLIN.net

The largest bodies of fresh water on earth are the five Great Lakes located along the border between the United States and Canada. Since the beginning of human dominance over other mammals on earth, we have inhabited the shores of fresh water lakes. A lake is a friendly place to live. It supplies drinking water, water to irrigate crops, and is a source of fish for the dinner table.

If land, such as Greenland, is too far north (or south) sub-zero temperatures keep the snow from melting. When this happens, snow build-up stores huge amounts of fresh water. The snow piles up year after year until it is hundreds of feet deep. During the summer the packed snow heats up and begins to flow as a glacier and may break off as an iceberg. Water stored in this way is of little use to us, although there has been some talk of towing icebergs to warm climes as a source of fresh water.

A large quantity of water is stored on high mountains such as the Rocky, Andes, and Himalayas where snow builds up during the winter and monsoon season. During the summer the melting snow flows off the mountains as streams that flow into an ocean or lake. Water draining off of mountains goes on to form some of the mightiest rivers on earth.

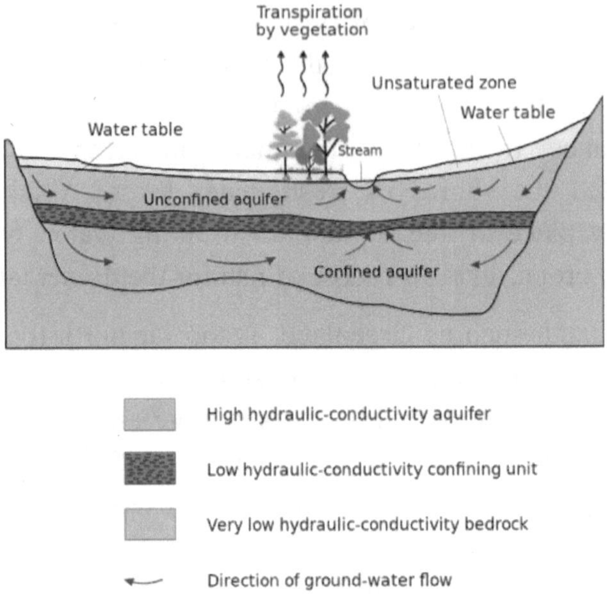

Figure 3.5 Typical Aquifer cross-section
Source: Wikipedia.com

In some places, such as deserts, streams dry up or disappear into underground aquifers. An aquifer is an underground layer of sand or porous rock where water is stored. We can get water from an aquifer by sinking a well into it. A huge aquifer supplies the city of Los Angeles with water. Rivers of water are sunk into it to keep up the water table so that there is never a shortage of fresh water. On the other hand, if the water table in an aquifer is higher than the surface of the surrounding land, the water will flow out as a spring. These springs or artesian wells are the source of the smallest brooks to the mighty rivers. Aquifers are nature's water storage tanks that keep the vital supply of fresh water flowing all year round.

Water is stored in a reservoir the shape of a lens beneath volcanic or lava islands. For example, the Hawaiian Island of Oahu is porous lava rock left over from a long extinct volcano. The water of the Pacific Ocean penetrates through the island at sea level. Water condensed by the cooler air above the mountains on Oahu falls on the porous lava rock and percolates through it until it reaches the ocean level. Being lighter than the salt water of the Pacific Ocean, the fresh water floats. The floating fresh water takes the form of a giant lens that is thickest in the middle of Oahu. Wells tap this constantly replenished, inexhaustible supply of sweet fresh water. Each of the larger Hawaiian Islands has its own lens of fresh water.

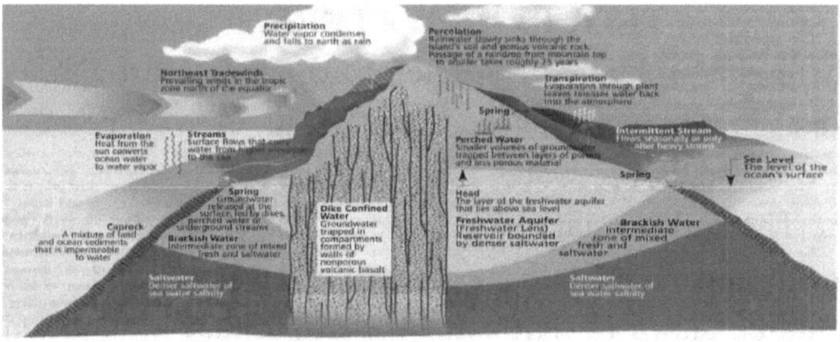

Figure 3.6: Schematic of Oahu's water source (not to scale)
Source: Board of Water Supply, City and County of Honolulu

THE UNEVEN DISTRIBUTION OF WATER

An uneven distribution of water upon the land is a problem. In some places on the earth too much rain creates jungles and swamps. In other places the lack of rain kills the vegetation and causes vast areas of the earth's surface to turn into barren deserts devoid of water. Such places are inhospitable to human habitation.

One of the ways used to solve the problem of uneven distribution of water is to transfer it from regions of over abundance to regions where water is scarce. One of the most prominent examples of the transfer of water is California. Water is diverted from the Owens, Feather, and Colorado Rivers into Southern California where the natural distribution of water within the region is insufficient to meet the needs of the population. The redistribution of water on such a grand scale requires the availability of financing as well as construction and engineering capabilities on an equally grand scale.

TROUBLED WATERS

There is a problem. The state of California, through its water distribution agencies, has been using up to 25 percent more Colorado River water than was agreed to in a 1929 accord. The other states that were party to the 1929 agreement raised a hue and cry and demanded that California be shorn of that extra water and that it be given to them instead. In 2003 California signed an agreement with the other users of Colorado River water that it would reduce the excess usage of Colorado River water to the 1929 authorized annual share of 4.4 million acre-feet within the next fourteen years.

Therefore, California has fourteen years to find a solution to a water problem in a region where there is no water available except that flowing in the Colorado River.

A water cut-off would adversely affect the Imperial Valley. The Imperial Valley is a “hot house” of rich irrigated farmland in the Colorado Desert of California, most of which lies below sea level. At the center of the Imperial Valley is the 350 square mile Salton Sea, the surface of which is 232 feet below sea level. The Salton Sea was formed when the Colorado River flooded into this below-sea-level depression between the years of 1904 and 1907.

Figure 3.7 The Imperial Valley
Source: National Atlas of the U.S.

The lesson is that regardless of the fact that California has spent billions of dollars to deliver water to where it is needed most; those efforts are scheduled to come to a halt soon unless another source of water can be found.

THE SEARCH FOR A DESALINATION SOLUTION

The Los Angeles Metropolitan Water District (MWD) says that it is already looking into ways to convert seawater into fresh water to satisfy future water needs to supply a burgeoning population. The process of desalinization requires the expenditure of a considerable amount of energy, and since expansion of non-polluting nuclear energy has been put on hold in the United States, fossil fuel will have to be burned. Generally speaking, there are two processes used to take salt out of seawater.

The first method requires that seawater be pumped under pressure through filters or membranes. The openings in the filter are small enough to keep the salt (NaCl) from passing through, while at the same time, are big enough to allow the water molecule (H20) to pass through. This procedure produces acceptable drinking water with about one percent salt. The second, and most used, method is to boil the seawater. The steam is piped through cooling coils and condenses to yield pure, distilled water. The oil rich countries in the arid Middle East have used the pressurized filter method of producing potable water from seawater for decades. It is also the method used by ships at sea.

The MWD has plans to build research plants and to subsidize the construction of as many as ten demonstration plants that can be brought on line to produce potable water at a future date. The problem is that it costs twice as much to put fresh, desalinized water at a water faucet as it does to put imported river water at the same faucet.

Desalinization projects are also ongoing in Texas and Florida. New Mexico has a project to desalinize brackish underground water.

OBAMA ADMINISTRATION SPEAKS OUT

President Obama's Secretary of Energy, Steven Chu, was quoted in the *Los Angeles Times* on February 4, 2009, as saying, "California farms, vineyards, and cities could vanish by the end of this century (ninety-one years), if Americans do not act to slow the advance of global warming. I do not think the American public has gripped in its gut what could happen. We're looking at a scenario where there is no more agriculture in California. I don't actually see how they can keep their cities going—either."

Obviously, Mr. Chu does not know that an unlimited quantity of water can be distilled by harnessing the sun's power. Furthermore, the means for doing it is nearby in the Mojave Desert at Kramer Junction, California.

SUN POWER

There is a better way than using the pressurized membrane system to desalinize salt water. That way is to use the free energy of the sun to produce pure distilled water in large quantities. The answer is to build parabola farms. The geometric definition of the parabola is a curve formed by the intersection of a plane parallel to the side of a right circular cone. This does not really concern us except for the amazing ability of the parabola to concentrate light rays from the sun so that they all converge at a single point that we call the "focal point."

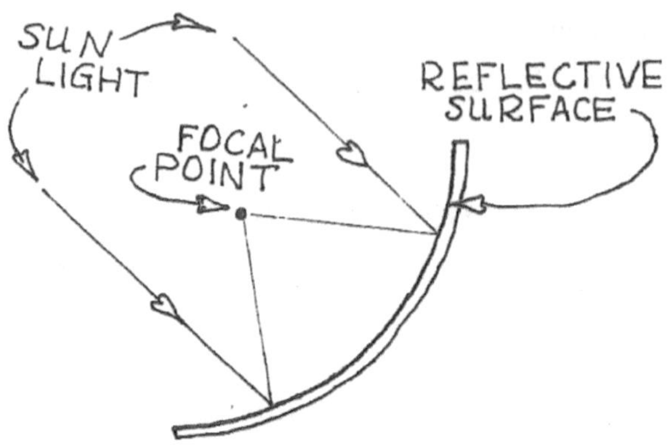

Figure 3.8. Schematic of a parabola, showing how sunlight is concentrated at the focal point

The most familiar types of parabolic configurations that come to mind are the "dishes" used to capture the satellite signal for our home televisions and the big "dishes" used by the telephone company to transmit telephone calls from one elevated point across a long line of sight distance to another elevated dish. These are parabolic dishes, of course, that emit signals from their focal points and send them out as parallel waves to be captured by another parabolic dish that concentrates them again at its focal point so that we can get a distortion-free digital television picture or a noise-free telephone call.

I chose two examples that use television and telephone waves because they illustrate a common usage of the parabola in our everyday lives.

A dish with a reflective parabolic surface can be used to concentrate sunlight or the rays of the sun at its focal point. Such a parabola can concentrate a considerable amount of energy (heat) at the focal point.

Of course, the quantity of heat produced by a parabolic surface depends upon the amount of sunshine available at a specific geographic location. The U.S. Department of Energy (DOE) says that the driest and hottest part of the United States—the southwest—receives about seven to eight kilowatt hours per square meter per day. Some places on the earth receive more than 9 kWh/m2/day of sunlight.

For decades the Department of Energy has conducted experiments in the southwest with the objective of converting sunlight into electricity. These expensive and sophisticated demonstration experiments are marginally effective for their intended purpose: production of electric power. Effective and reliable power production demands that the process provide electric power continuously. In a word, these experiments are demonstration power plants that require fossil fuel generators to kick-in to carry them through the night.

A lot of electricity can be produced from the sun. The DOE has calculated that enough electric power to supply the needs of the entire country could be generated by covering about 9 percent of Nevada, a plot of land 100 miles on a side, with a parabolic trough system.

THE PARABOLIC TROUGH

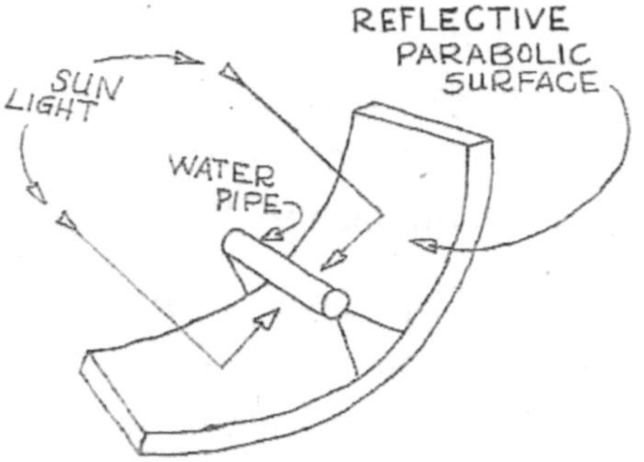

Figure 3.9. Schematic of a Parabolic Trough System.
Shows sunlight focused on a water pipe at the focal line.

The parabolic trough system focuses the sun's rays by bouncing them off of the reflective surface of a parabolic trough onto a pipe running along the focal line of the trough.

Although the DOE has built several solar power systems, the one that offers the best design for commercial operation is the trough system. The concentrated energy of the sunlight heats oil flowing through the receiver pipe, and the heat energy is then used to generate electricity in a conventional steam generator. The collector field comprises many troughs placed end-to-end with a continuous receiver pipe inside the troughs along the focal line.

Figure 3.10 Solar Electric Generating System (SEGS) troughs at Kramer Junction, California.
Source: Sandia Laboratory

To configure a large parabolic trough field many parallel rows of long trough systems are aligned on a north-south axis. This configuration enables the single-axis troughs to track the sun from east to west during the day to ensure that the sun is continuously focused on the receiver pipe. Individual trough systems can generate about eighty megawatts of electricity.

Figure 3.11 Aerial Photograph of the Solar Electric Generating System (SEGS) at Kramer Junction, California.
Source: Sandia Laboratory

SOLVING THE CALIFORNIA WATER SHORTAGE

The Department of Energy (DOE) does not have a project to use a parabolic trough collector field to turn seawater into fresh water by boiling it and condensing the steam to get distilled water. It is certain that such a system is exactly what the Metropolitan Water District (MWD) of Southern California needs to make sure its eighteen million customers are assured a plentiful supply of fresh water. Such a system should be able to produce more than enough water to take care of future population, agricultural, and industrial increases in Southern California.

Excess water would be stored in aquifers, lakes, and tank farms to be tapped as needed. The big advantage of such a system is that it would be environmentally friendly. After a parabolic trough system is in place and paid for, the abundant sunlight of the arid southwest would be used to supply a basic need of the population—free fresh water from sun power.

I recommend use of the existing Department of Energy (DOE) parabolic trough design in which oil is heated as it flows through the focal pipes. The hot oil flows into an insulated heating chamber where sea water is flash boiled. The steam is processed through condensers to produce pure distilled water. It would be easy to sweep the salt out of the bottom of the heating chamber.

Instead of spending $100s of millions to fund construction of expensive fossil fuel research and demonstration plants over the next decade, a much better solution would be for MWD to use the expertise of the

DOE to build parabolic trough fields, or to make use of the existing SEGS facility at Kramer Junction, California.

The advantage of these systems cannot be over-emphasized. They are proven and tested technology. They do not pollute. They save energy by using no fossil fuels. They are simple and easy to use, and are low maintenance. They are environmentally friendly. They are safe.

A parabolic trough system would be an ideal solution for the farmers of the Imperial Valley where the bureaucrats plan to reduce the irrigation water available from the Colorado River because of a 1929 water sharing agreement.

Of course, water management or distribution would not be free. In order to minimize overhead costs, the water produced by solar trough farms should be merged into other local water supply systems.

COMMUNICATION WITH THE GOVERNOR

OFFICE OF THE GOVERNOR

June 17, 2008

Mr. Daniel Myers Jr.
92-1210 Uahanai Place
Kapolei, Hawaii 96707

Dear Mr. Myers,

Thank you for taking the time to send me a copy of Food Water Energy and Human Survival.

I am working hard to improve California's future and make progress in a difficult budget year. We've tackled politically risky things that no one wanted to touch in the past, and to me, that's progress. We now have the potential to make a positive difference in the lives of future generations by confronting problems that have been put off for many years: our budget, health care, education and our state's infrastructure.

I appreciate your thinking of ways we can improve our great state. Because of citizens like you, we will get California back on track. I believe we will turn today's temporary problems into permanent victories for the people of our great Golden State.

Sincerely,

Arnold Schwarzenegger

GOVERNOR ARNOLD SCHWARZENEGGER • SACRAMENTO, CALIFORNIA 95814 • (916) 445-2841

June 6, 2008

Governor Arnold Schwarzenegger
State Capitol Building
Sacramento, California 95814

RE: California Water Shortage, Encl. 1.

If you took action today, California could have *billions* of gallons of fresh water within a short few years.

The answer is to use *solar* power to produce an *unlimited* amount of water by boiling Pacific Ocean water and condensing the steam to get distilled water.

I have outlined the way to do this in my book FOOD WATER ENERGY and HUMAN SURVIVAL (Encl. 2). Please refer to the chapter on WATER beginning on page 81.

I wrote a letter to your Secretary of Resources, Mike Chrisman, on February 5, 2008 and explained to him how to produce fresh water from seawater, Encl. 3.

I hope you will act now to start the State of California on the path to solving its water shortage.

Sincerely,

Daniel J. Myers, Jr.
92-1210 Uahanai Place
Kapolei, HI 96707-1562
Phone: (808) 672-5325
Email: danmyersjr@yahoo.com
Encl.1: Article WSJ: Governor Declares Statewide Drought.

U.S. SOLUTIONS FOR THE IMPERIAL VALLEY

The first location that I would recommend for a new parabolic trough farm would be along the Pacific Ocean near San Diego. It would produce more than enough to make up all the anticipated water shortages of Southern California. Furthermore, the construction of a water pipe eastward over the mountains for about 150 miles to the Imperial Valley would take care of the impending cut off of water from the Colorado River. This would be a relatively expensive project but has the advantage that the whole operation would be within the borders of the United States.

A second, less expensive and more elegant solution would be to use the Solar Electric Generating System already in place at Kramer Junction. This would be easy to do if ocean water was piped into Kramer Junction from the Pacific Ocean. Of course, the distilled water would be piped across the mountains to the Imperial Valley.

MEXICIAN SOLUTION FOR THE IMPERIAL VALLEY

A third option would be to place a parabolic trough farm on the Gulf of California in Mexico and pump the water north about 150 miles directly into the Imperial Valley. A better route would be to pipe the water about 100 miles north to the All American Canal. The canal is fifty miles nearer to the Gulf of California and would take advantage of the existing water distribution system. The All American Canal is an eighty-mile-long canal that waters the rich Imperial Valley. Its water comes out of the Colorado River near Yuma, Arizona, and from there flows westward into the Imperial Valley. (Please refer to Figure 3.7) Water could be stored in a tank farm, or an aquifer, or poured directly into the All American Canal to replace Colorado River water lost to other states by California when the federal government begins to enforce the 1929 water agreement.

This arrangement would require that the United States and Mexico sign an agreement to permit construction of a pipeline, a parabolic trough farm, and use of seawater from the Gulf of California. The advantage of this option is that the land between the Gulf of California and the Imperial Valley is low-lying and rather flat, making construction and maintenance less expensive. Another advantage of this placement is that there are more hours of intensive sunshine along the Gulf of California than along the Pacific Coast near San Diego, or Kramer Junction California

Of course, the United States should offer to sell the parabolic trough system to the Mexicans so that the United States could build more of them along the arid west coast of Mexico and around the Gulf of California.

DROUGHT IN EUROPE

The summer of 2005 was a weather disaster for Portugal, Spain, France, and Italy caught in a heat wave that caused dwindling water supply with crops withering in fields turned to dust. Portugal and Spain suffered the worst drought since records began to be kept in 1940. In France water levels were the lowest since 1976. Parched conditions stretched from North Africa to Paris, causing billions of dollars in damage as crops withered, rivers dried up and pastures turned to dust.

The drought crippled crop production in Spain, Portugal, and North Africa. The European Union said, "The Iberian Peninsula faced the worst conditions in the last 30 years. The situation is critical."

Additionally, the hot dry weather in southern France spawned swarms of locally-hatched locust that ate their way through crops and gardens.

FRESH WATER FOR EUROPE

Several well-placed parabolic trough farms along the coast of the Mediterranean Sea in North Africa, Spain, Portugal, France, and Italy would provide an abundance of fresh water for crop, human, and industrial consumption for a fraction of the cost of a drought. Long sunny days that prevail around the perimeter of the Mediterranean in Europe, Africa, and Asia makes the climate ideal for water production using the Solar Electric Generating System (SEGS).

MAKING THE DESERTS BLOOM

The great deserts of Australia, Asia, Africa, and the Americas could be made to bloom if parabolic trough farms were built along their coasts to convert salt water out of the oceans into fresh distilled water. An inexhaustible supply of solar power generated fresh water could be stored under and above ground until it is needed for people, industry, or to irrigate crops. Such a supply of water would open up millions of acres of rich farmland and relieve population pressure on cities. We can imagine the wealth and well being that would accrue to the Arabian Peninsula in southwest Asia or the Sahara Desert in northern Africa, for example, if an unlimited supply of fresh water was available for household consumption, agriculture, and industry. With the use of parabolic trough technology fresh water could be produced on a macro scale—billions of gallons.

The deserts would bloom.

In the not-too-distant future, the wealth of nations will

be counted by the miles of pipe lines flowing with fresh water instead of oil. A new era of prosperity and peaceful development is within our grasp.

Only the water is missing.

A SOLAR STOVE FOR THE FAMILY

There is a need to adapt the solar concentrating power of the parabola for use on a micro scale as a source of heating by individuals or families, especially in the developing countries of the world.

THE ASIAN BROWN CLOUD

The urgency of the need for a pollution-free heating source adapted to use at the family level was highlighted in a study by 200 scientists between 1995 and 2000. The Reuters news service reported that the study under the auspices of the United Nations Environment Program used data from ships, planes and satellites to study the two-mile-thick cloud of pollution that hangs over South Asia. It is causing premature deaths of a half-million people in India each year. "The grimy cocktail of ash, soot, acids and other damaging airborne particles is as much the result of low-tech polluters like wood- and dung-burning stoves, cooking fires and forest clearing as it is of dirty industries," the study found. The so-called "Asian Brown Cloud" appears to have altered the region's monsoon rains—increasing rainfall and flooding in Bangladesh, Nepal, and northeastern India, while cutting back needed seasonal rain in Pakistan and northwestern India.

"When you think about air pollution, many people

think of industry and fossil fuels as the only causes," co-author Paul Crutzen, a scientist at the Max-Planck Institute for Chemistry in Mainz, Germany, told a news conference on August 12, 2002, in London. "Often ignored," he said was, "biomass burning, including forest fires and the burning of vegetation to clear land or to warm the homes of poor people."

Veerabhadran Ramanathan of the Scripps Howard Institute of Oceanography in La Jolla, California, one of the report's authors, said "the extent of the sunlight loss was a major surprise." We've been looking at environmental issues for the last several decades, yet Asian haze came as a major surprise to us." He said, "We don't know how many more surprises we will find."

For many years, scientists believed only lighter greenhouse gases, such as carbon dioxide that is produced from burning fossil fuels such as gasoline, diesel, and oil, were global in reach and effect. They now say that microscopic, suspended particles of pollutants, generically called aerosols by atmospheric scientists, also travel the globe.

The magnitude of the problem was a surprise to officials at the United Nations. The "Asian brown cloud" reduces the amount of sunlight reaching the surface of the earth by 10 to 15 percent, cooling the land and water while heating the atmosphere. Floods, drought, sunlight reduction and acid rain all can hurt agricultural yields, with the report indicating the pollution may be cutting India's winter rice harvest by as much as 10 percent.

Klaus Toefler, executive director of the UN Environment Program, said scientists and policy makers, "should avoid making premature assessments, but should start trying to cut pollution by introducing more efficient heating stoves in developing counties and turning to solar power and other clean sources of energy."

PROBLEMS WITH SAFE DRINKING WATER

At the World Summit on Sustainable Development that was held in Johannesburg, South Africa, from August 26 to September 3, 2002, the need for safe drinking water for people who live in the poor nations of the world was a subject of intense discussion. The Inter Press Service reported that the Water Supply and Sanitation Council pointed out that one of the biggest errors of international development in the last fifty years has been the failure to make clean water, safe sanitation, and knowledge about hygiene availability in the third world.

Illnesses related to a lack of clean water and sanitation cause about five million premature deaths a year with more than two million young children dying of diseases like cholera and dysentery each year. "Overall it can be estimated that at any one time almost half of the people in lesser developed countries are suffering from diseases associated with a lack of clean water and sanitation," reported the Council.

Most nations of the world, especially in the third world, do not have the ability to match the California example of gathering water from hundreds of miles away. They are not able to manage, finance, process, and deliver water through underground pipes as safe drinking water to millions of families. The magnitude of the problem facing the third world nations can be understood by the fact that an estimated two billion people or a third of the population of the world, most of which live in under developed countries, lack access to safe drinking water and sanitation.

PARABOLIC STOVE TO BOIL WATER AND CLEAR THE AIR

There is an inexpensive solution that can go a long way in helping to solve the sanitation problem. The solution is to use the solar concentrating power of the parabola (please see Figure 3.8) to provide the heat needed for boiling water and cooking food for the family table. In order to bring the magic of solar heat to the world, inexpensive solar collectors (stoves) should be manufactured and sold or given to families that need them. The solar stove would be light in weight, portable and with a reflective parabolic surface that could be pivoted by hand so as to focus the rays of the sun on the focal point or "hot spot" where the cooking or boiling would take place.

The parabolic stove already exists. The *National Geographic* Magazine reports in its January 2002 issue on an expedition in September-October 2000 to explore the Gobi Desert of northern China.

On the third day members of the expedition arrived at the home of a seventy-two-year-old woman named Diudiu who welcomed them to spend the next few days with her.

In preparation for afternoon tea, she went inside her house and filled a tea kettle with water from a small cistern, then walked outside to a mirrored solar collector the size of a large TV satellite dish.

At the center of the dish, where the rays of the sun were to be focused, Diudiu snapped the kettle into an iron fitting and pivoted the dish to face the afternoon sun. In seconds the kettle was smoking. Within three minutes, the water was boiling furiously.

"I sold hair from my camels and sheep to buy this on the outside," she says turning the mirrored face of the dish from the sun to retrieve the kettle. "It keeps me from having a fire going all day."

Let me repeat: "It keeps me from having a fire going all day." Diudiu has set an example that needs to be duplicated millions of times, especially throughout the third world. If the governments of south Asia would start to make solar stoves available to the people living under the "Asian brown cloud," the air would clear up.

Figure 3.12. A Chinese woman heating water on her solar stove. (By permission of George Steinmetz,

photographer.)

I recommend that stoves be mass produced at a fraction of the cost that Diudiu paid for her large parabolic stove. Typically, a solar stove would have a parabolic disc of, say, several feet in diameter. It would be easy to assemble, disassemble, and move.

At the present time, poor people living under that dirty cloud are denuding the landscape to scrounge wood, grass, weeds, leaves and dung for cooking and heating. The forests have been depleted so much that in many areas women must walk miles every day to collect and carry firewood. The use of solar stoves would allow the forests to grow again.

SOLAR TROUGH STOVE

The solar stove as a source of heat has application far beyond that which might be envisioned strictly for third world use. In the more developed nations of the first world, families could install a solar trough, say six to twelve feet long, synchronized to move with the sun as a cost-free way of heating their homes. A solar trough concentrates the rays of the sun along a focal line rather than at a focal point (please see Figure 3.9). The trough configuration generates more useable energy than a solar stove when the sun is shining. The solar trough would be especially useful in northern and southern latitudes during winter months when the direct rays of the sun are diminished as the earth "wobbles" in its annual orbit around the sun. In cold weather, the parabolic trough solar collector pointed at the sun to receive maximum solar density is an efficient and useful source of heat as long as the sun is shinning.

FAMILY APPLICATION

A solar trough installed on the roof of a house using solar trough heating would be equipped with a closed loop water circulating system. Water would pass through the heating pipe at the focal line of the solar trough and pumped for storage into a large thermos tank, say 1000 gallons or more, for general household use. The installation of a double-walled thermos tank would serve the obvious purpose keeping the water hot when the sun is not shining. If desired, the closed water loop could include radiators to provide heat throughout the house or for specific rooms. A fossil fuel or electric heating system would back up the solar-powered heating system.

Those who wanted to take full advantage of the free solar energy provided by the sun could install an air conditioning system using the solar trough as its heat source. The air conditioning unit would consist of a closed loop that used ammonia (NH3) as the refrigerant in an ammonia cooling cycle.

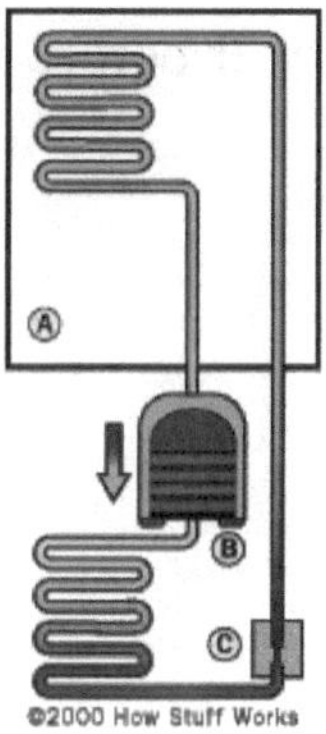

Figure 3.13. Ammonia refrigerator cycle
Source: How stuff works

The ammonia cooling cycle is old, well-developed technology that has been around since the 19th century that used a gas or candle flame as the energy (heat) source. An electric heating system would kick in to provide heat for the unit when the sun was not shining.

COMMERCIAL AND PUBLIC BUILDING APPLICATION

The savings that would be realized by installing solar trough heating and air conditioning systems on the top of large commercial and public buildings can not be over emphasized. Think about it: free, environmentally friendly solar heating and cooling when the sun shines. Regular heating and air conditioning systems would supplement the solar system and kick in at temperatures set by a thermostat.

The installation of parabolic solar farms to produce massive amounts of fresh water combined with the use of solar stoves by families and businesses would produce wealth and well being never before imagined.

Let us do it.

CHAPTER 4: ENERGY

Energy, Energy. All is Energy!

THE VORACIOUS AND HUNGRY ENERGY MACHINE

Food and water are the fuels necessary to sustain life. That fuel is converted into energy and is used by all living organisms to stay alive. It does not matter whether the consumer is a one-cell ameba or a human. Our bodies must have the body of other organisms to eat if we are to survive. All animals eat other animals or plants to stay alive. The human body is a voracious and hungry energy producing machine that devours the bodies of other animals or plants in order to live.

Cruel, perhaps, by some definitions, but, nevertheless, true. It is an immutable law imposed upon us by nature. This immutable law imposes a second law upon the animal kingdom, namely, the law of the survival of the fittest. Survival of the fittest means simply that priorities are established for those who eat other organisms and those that are eaten; that is, the predator and the prey; which brings us full circle to the survival of the fittest. Natural selection has succeeded in making it possible for both predator and prey to survive and prosper.

Vegetarians argue that it is inhumane or sinful to eat other animals. They say that we should only eat plants. This is a silly argument because plants are living organisms. Plants and animals are eaten as fuel by some animals such as humans. The living cells of plants give up their energy to some animals that live exclusively on them, the so-called ruminants; for example, cows, horses, and sheep.

The living plant is a complex and cosmic organism that takes its sustenance from the earth and from the sun through a process known as photosynthesis. Thus, the so-called lowly plant is the giver of most of the energy on earth, from the decayed plants we pump out of the ground as oil to the golden fields of grain waiting to be harvested. Plants are truly cosmic denizens that reach out and gather the warmth of the sun to give us an inexhaustible supply of food on earth.

THE UNIQUE HUMAN

As noted above, the human body is a hard working engine that continually consumes food and water to generate energy every second of our earthly existence. Our body is a natural work of art. Humans are unique in all of nature in that we can think, talk and create. We have created civilizations, empires, and religions. We are artists, musicians, engineers, scientists, poets, and much more.

Through the sheer power of our minds, we have created a sophisticated modern world that, like the human body, depends upon the consumption of energy. The current energy of choice is fossil fuel and nuclear fuel. We must take care that this remarkable creation that flows from the mind of man—modern civilization—is kept alive and spread throughout the planet for the greater good of all.

DEATH OF THE BODY

If, however, we consider only the body and not the mind, we see at once how dependent we are on a continuous supply of food and water. If either or both is denied, we die. We will all die, of course, whether or not we have food and water. The body grows old and wears out. Death is the fate of all living organisms, large and small, plant and animal. When we die we return to the earth from whence we came. Our body is dead. It will rot or burn and go back to Mother Earth.

Is that all there is?

CONCEPT OF THE ETERNAL SOUL

Must we return to dust after such a short, hard life of misery, toil, and worry? In truth, no one knows; many believe that humans have souls that are eternal and will forever occupy a spiritual region beyond the earthly perception of the living. This concept of life after death is the basic foundation of all religions. The religious believe that the human spirit will reside in a heavenly paradise forever or return to earth periodically as a reincarnated living creature. Belief in the afterlife, however, is a matter of personal faith and is beyond the scope of this book.

STARDUST

Whether there is or is not a spiritual life after death, it is certain that we will exist forever. Some of the atoms that make up the universe were ours during our brief sojourn here on earth.

The same atoms that breathed life into you and me

exist everywhere throughout the billions of stars in billions of galaxies. If, indeed, there is a soul that lives beyond our earthly death, then it can truly be said that we are a part of the infinite universe. When the atoms that make up our bodies return to earth, they return to the universe from whence they came.

In fact, the living body is an array of billions of atoms. It is a virtual universe within itself. It is a universe that will die and return to the stars when the breath of life no longer sustains our mortal beings on this earth. At death, we exchange the atoms that make up our earthly bodies for an infinite eternal existence in the infinite universe.

We are cosmic. We are stardust.

ATOMS: BUILDING BLOCKS OF THE UNIVERSE

Atoms are the keys to understanding the universe. Sir Isaac Newton, the greatest scientific mind who ever lived, wrote the universal laws of force, motion, and attraction and showed their relationship to each other by inventing calculus to prove it with his experiments. He taught us that all matter is attracted to all other matter by a pull we call gravity. For example, the force of gravity acting between you and the earth keeps your feet firmly planted upon the ground. We measure the force of gravity every time we step on the scales. The pull of gravity keeps our atmosphere and oceans from drifting away from earth into outer space. The pull of gravity keeps the moon rotating around the earth and the earth rotating around the sun and the sun rotating around our galaxy, the Milky Way.

THE SUN

Our sun, an average size star, is an awesome energy producing machine that gives life to us all. The sun is made up of billions and billions of atoms, mostly hydrogen (1H1), that are pulled together by the force of gravity. The many billions of atoms that make up the mass of the sun create a gravitational force that pulls them toward the center of the sun. The magnitude of the pull is so great that it is almost impossible for us to imagine. The pressure at the center of the sun is said to be about 340 billion times the air pressure at sea level. The temperature at the center of the sun is said to about 15 million degrees Centigrade (27,000,000□□Fahrenheit). Under these conditions of extreme pressure and temperature, the hydrogen (1H1) nuclei are fused together to make an alpha particle or Helium (4He2) nucleus.

$$4\ 1H1 \rightarrow 4HE2$$

In the process of going from four H nuclei to one He nucleus, it has been experimentally determined that there is 0.7 percent less mass in the resulting helium nucleus than the original 4 hydrogen nuclei. The missing mass is converted into energy, called electromagnetic radiation, that migrates to the sun's surface in approximately a million years. From the surface of the sun the energy from its internal fusion furnace is radiated into space as a broad spectrum of electromagnetic radiation that includes sunlight.

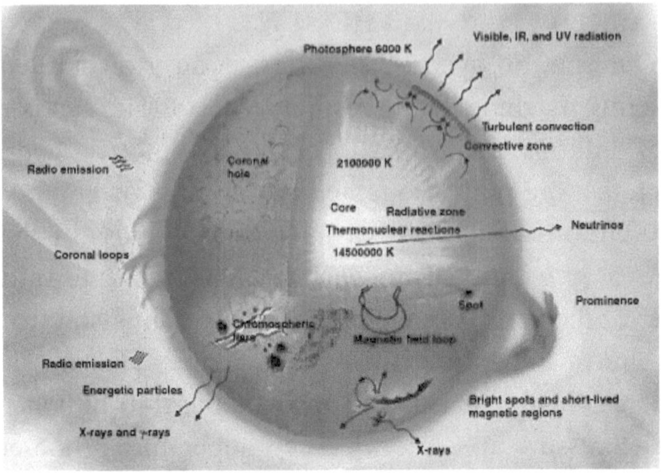

Figure 4.1. NASA Cross-section of the Sun
Source: Wikipedia.com

The sun is burning up. It is estimated that 600 million tons of hydrogen is converted to helium and that 4 million tons of energy is released every second. Not to worry. The sun is about 4.5 billion years old and has enough fuel to last for 5 billion more years. The sun will continue to send its life-giving rays out to us for a long, long time.

DEATH OF THE SUN

What will happen to the sun when it has fused all of its hydrogen into helium? The internal fusion furnace of the sun will start to die down, causing the exterior of the sun to expand and cool. The surface of the sun will expand into a big fire ball that will incinerate the planets, life on earth, and convert its heavier elements back into gases. With the hydrogen gone, the core collapses upon itself as the helium tries to fuse into carbon (12C6); an atom that is too heavy for the gravity of our sun to fuse. The new diameter of the sun will be about the same as that of earth. Meanwhile, the outer mass of the sun drifts off into space as stardust. The sun continues to glow and becomes what we call a white dwarf. After billions of years the white dwarf will cool down, stop glowing, and become a black dwarf wandering in space.

When our sun dies it will not give off enough stardust to create another star. The stardust will remain in space as cosmic clouds. A star can be born if one of the billions of galaxies in the universe collides with our galaxy—the Milky Way. If the Milky Way collided with another galaxy, the stardust of our dead sun could mix with the stardust of the passing galaxy to give birth to a new star.

A STAR IS BORN

Stars are born within a galaxy when a very large star explodes and gives off enough hydrogen to mix with existing cosmic clouds to provide the hydrogen to make another star. Big star explosions are called Super Novae. The energy of the explosion liberates hydrogen by breaking up helium atoms through the process of fission. As a large number hydrogen nuclei come together, they start to pull on each other (attract), just as we were taught by Sir Isaac Newton. The mass of gas grows slowly by sweeping in more cosmic clouds as the force of gravity pulls the entire mass toward its center. The atoms are compacted by the pull of gravity until the pressure and temperature is high enough to cause the hydrogen atoms to fuse into helium. The loss of mass that takes place during fusion is radiated into space and a new star is born.

INFINITY

This rhythm of death and rebirth of stars is a cosmic cycle that has gone on for billions of years and will continue for billions more. Unto infinity, if you can imagine that. As certain as our birth is and our death will be, after our short lifetime on this earth, we will rejoin the cosmos from whence we came as stardust.

FUSION ON EARTH

We know that the universe is a mighty energy-producing fusion furnace that converts the mass of the billions upon billions of stars into electromagnetic energy that is radiated throughout the cosmos. Suppose, however, that we could build a practical fusion furnace on earth that is big enough to heat steam to run an electrical generator.

To be successful such a furnace would have to be able to create temperature and pressure similar to that in the interior of the sun in order to make hydrogen nuclei fuse into helium. Nuclear physicists have been able to cause hydrogen to fuse by using atom bombs as primers to duplicate the temperatures and pressure of the sun in hydrogen bombs. Such bombs have released megatons of energy into the atmosphere and can wipe out a whole city in a flash. Obviously, this is not the solution we are looking for to harness the fusion of hydrogen nuclei as a source of usable energy.

PROJECT SHERWOOD

If it is to be useful, the furnace must be able to fuse hydrogen nuclei in a sustained and controlled reaction that gives up the surplus of energy created by the conversion of mass into energy. In the middle of the last century, United States scientists were so sure that a fusion furnace could be built on earth that the program was classified top secret and called Project Sherwood.

They were wrong.

Since that time the United States has spent billions of dollars at dozens of universities and laboratories in an effort to build a usable fusion furnace.

All have failed.

THE PROBLEM WITH FUSION FURNACES

Fusion occurs when there is enough energy available to cause hydrogen nuclei to come together to form a new and larger element, namely helium. Hydrogen is the perfect choice for fusion because it is the first element in the periodic table of elements with one proton (1p+1) that, by convention, carries a positive charge of plus one. The hydrogen atom (1H1) can, however, be modified by accepting into its nucleus one or two neutrons (1n0). Neutrons have a zero charge, therefore, the charge on the hydrogen atom remains at plus one. These modified hydrogen atoms are called deuterium (2H1) and tritium (3H1) because they have different atomic weights but remain hydrogen because the number of protons in the atom remains the same. They are isotopes of hydrogen. An isotope is an atom that has a heavier or lighter atomic weight than normal but keeps the number of protons in its nucleus unchanged.

The problems encountered in building a fusion furnace

are challenging. For example, the temperature and pressure that must be maintained is the same as that in the interior of the sun. At this temperature and pressure a material containment vessel is impractical because it would turn into hot gaseous plasma. The accepted solution is to create a strong "magnetic containment vessel" inside a powerful vacuum container as the fusion furnace. But the magnetically contained fusion furnace does not work because the hydrogen nuclei leak out along its lines of magnetic force.

The fusion furnace is very desirable as a source of energy because it is "clean." This means that little harmful radioactivity is given off during fusion and that there are few or no radioactive byproducts.

It is, of course, impossible to know whether or not someone will someday succeed in making a fusion furnace, but the prospect is not bright at this time. Let us hope that someday some genius will succeed.

FISSION

We do not have fusion furnaces but we do have *fission* furnaces, or nuclear power plants, that drive electrical generators. Fission is the opposite of the fusion process and since our search for a fusion furnace has failed, we must look elsewhere.

For an answer let us begin by examining the periodic table of elements. There are 92 naturally occurring elements on earth. Each is identified by its atomic number, that is, hydrogen (H) is number 1 and uranium (U) is number 92 in the periodic table of elements. In recent years the periodic table has been extended beyond the 92 naturally occurring elements. These are the so-called transuranic elements, such as plutonium, that were created in a laboratory by nuclear physicists.

A closer examination of the first and last elements shows that as we proceed up the periodic table of elements, the number of neutrons in each nucleus increases rapidly. For example, naturally occurring hydrogen has zero neutrons whereas the most stable form of uranium has 238 protons and neutrons and is written as (238U92). By subtracting 92 from 238, it can be determined immediately that there are 146 neutrons in the most stable and naturally occurring uranium atom. Also, a number of isotopes of uranium can be created when a neutron is added to or expelled from its nucleus.

NUCLEAR INSTABILITY

The large number of neutrons that are found in the nucleus of heavy atoms cause them to be unstable. This instability causes them to split apart, or fission, into two or more smaller elements. In so doing the mass of the fission products are less than the original mass of the heavy element before fission. The mass that is lost in the process of fission is converted into energy and released as electromagnetic radiation. Once the process of producing energy through fission was understood, the challenge was to harness the energy and use it to operate a practical fission furnace capable of turning an electrical generator. The fission byproducts are radioactive and can pose a threat to life as compared to fusion products.

FISSION ON EARTH

Most of us are aware of nuclear energy because an atom bomb with an explosive force equal to 16,000 tons of TNT was exploded over the city of Hiroshima, Japan on August 6, 1945. When a second atomic bomb was exploded a few days later over Nagasaki the Japanese Empire sued the Allied Forces for peace, whereupon, World War II came to an end.

The atomic explosion over Hiroshima was made possible by the discovery that the uranium-235 isotope (235U92) would accept a slow moving neutron (1n0) into its nucleus and split apart with the release of 2 or 3 more slow moving neutrons that could then be taken in by other nearby uranium-235 atoms that would split again, and so on. This process is called a chain reaction. The energy that is released by fission comes from the fact that the original mass of the uranium-235 atom is heavier than the 2 or more new elements created as a result of the atom splitting apart. The loss of mass is released as electromagnetic radiation. Uranium-238 (238U92), which is the common form of uranium found in nature, has among its atoms only 0.7 percent of the fissionable uranium-235 isotope. Therefore, it was necessary to refine the uranium found in the earth's crust until the uranium-235 content was at a density above 90 percent in order to sustain a chain reaction. Therefore, if its density is above 90 percent and the mass is big enough (a critical mass), uranium-235 will sustain a chain reaction. The energy of fission is released almost instantaneously as a very destructive explosion that gives off blast, heat, and radioactive fallout.

A manmade atom, plutonium (239Pu94), is also fissionable with the release of about 3 neutrons per fission. The atomic bomb exploded over Nagasaki was a plutonium bomb. Plutonium is a transuranic element that is manufactured by exposing uranium-238 in a so-called breeder reactor.

$$238U92 + 1p{+}1 \rightarrow 239Np93 + 1p{+}1 \rightarrow 239Pu94$$

In this process, uranium-238 is changed into neptunium (239Np93) that is changed into plutonium-239. By transforming non-fissionable uranium-238 into fissionable plutonium-239, the reactor is said to “breed” nuclear fuel. By breeding fissionable plutonium-239 from non-fissionable uranium-238, the amount of nuclear fuel available on earth is increased by a factor of more than 100. Keep in mind that only .007 or 0.7 percent of naturally occurring uranium-235 mixed in with uranium 238 is available as a fissionable fuel; whereas, easily produced plutonium is excellent for use in atomic and hydrogen bombs as well as in fission furnaces or nuclear reactors.

NUCLEAR REACTORS

Shortly after the end of World War II, nuclear scientists began experimenting with reactors that could release energy a little at a time over a long period of time. As it turned out the fission furnace, unlike the fusion furnace, was very simple to build. Not only that, it could be controlled safely and operate for years each time it was reloaded with fissionable fuel rods.

The key to success was to make the fuel rods out of only about 3 percent fissionable fuel and dunk the whole reactor core into a big tub of water called the reactor containment vessel. The remarkable success of the fission furnace depended upon controlling the heat produced by the reactor core. The core can be made of as many as 40,000 fuel rods which are kept "cold" by inserting cadmium (112Cd48) control rods into it. Cadmium is a good absorber of neutrons and, therefore, keeps an uncontrolled chain reaction from taking place.

The control rods are pulled out of the core just enough for the heat of fission to turn water into steam, but not enough to cause the core to melt down. The two hydrogen atoms in the water molecule act as moderators to slow down and hold neutrons released outside the core from leaking out of the containment vessel. New elements created at fission, the so-called daughter products, are retained in the rods until the nuclear fuel is used up. New fuel rods are put in about every three years.

Nuclear power plants operate on the same principle as coal- and oil-fired power plants. Water is heated to produce high pressure steam that turns turbines that drive electrical power generators.

There are two types of nuclear power plants in operation today. They are the Pressurized Water Reactor (PWR) and the Boiling Water Reactor (BWR).

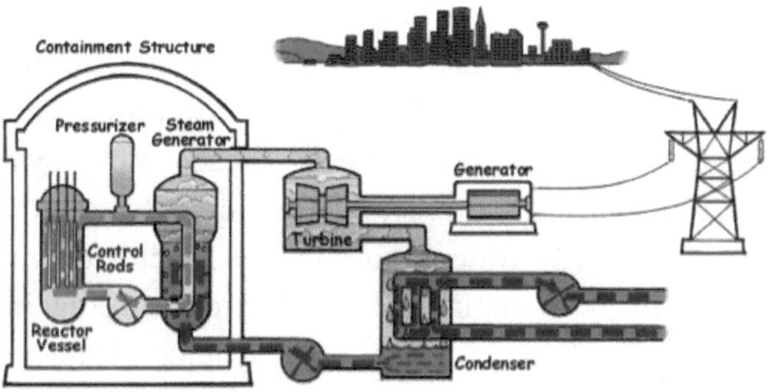

Figure 4.2. Schematic of Pressurized Water Reactor (PWR)

The PWR keeps water under pressure so that it heats, but does not boil. Water from the reactor and water in the steam generator that is turned into steam never mix. In this way, most of the radioactivity stays in the reactor area.

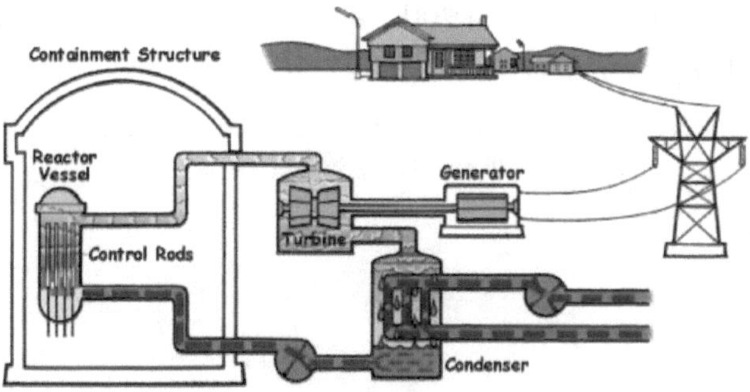

Figure 4.3. Schematic of a Boiling Water Reactor (BWR)

The BWR boils water in the reactor vessel. In both types, water is converted into steam, and then recycled back into water by the condenser, to be used again in the boiling process.

Both Schematics are Nuclear Regulatory Commission Graphics.

SAFETY

Since radioactive materials can be dangerous, nuclear power plants have many safety systems to protect workers, the public, and the environment. These safety systems include (1) shutting the reactor down quickly to stop the fission process, (2) systems to cool the reactor down and carry heat away from it, and (3) barriers to contain the radioactivity and prevent it from escaping. Radioactive materials, if not handled properly, can damage human cells or even cause cancer over a period of time.

NUCLEAR POWER PLANT CONSTRUCTION

In the 1950s nuclear power plants began producing electricity commercially. Today the world produces as much electricity from nuclear power as it did from all sources combined in 1960, according to the Uranium Information Center (UIC).

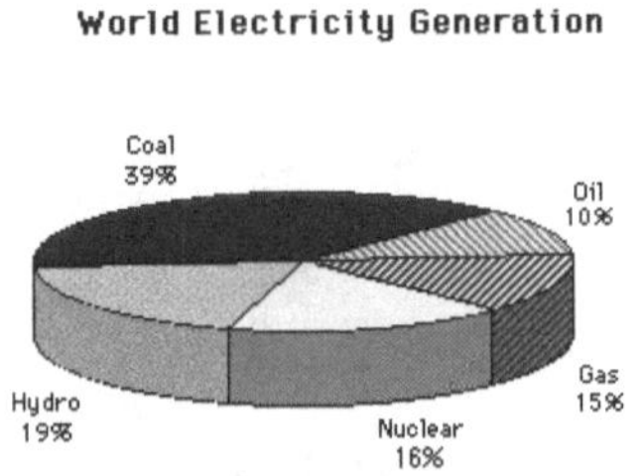

Figure 4.4. World Electricity Generation

The UIC says there are 440 commercial nuclear power plants in the world producing 16 percent of the world's needs for electricity as of March 2004. 30 power reactors are under construction with a similar number firmly planned for construction, mostly in China, Japan and South Korea.

What a remarkable achievement!

The mind of man discovered the secret of how to put uranium-235 at the service of all by harnessing the heat (energy) of fission for the production of electricity. Uranium-238 was transformed into plutonium-239, an excellent fissionable metal. Furthermore, the availability of uranium-238 is, for all practicable purposes, unlimited.

Amazing!

U.S. SHUTS DOWN NUCLEAR POWER PRODUCTION

Irrational fear took hold in the United States. The world had arrived at that sweet spot in history where everyone could enjoy electricity produced by a nuclear fission furnace. This is a dream that could be true, but it was not to be. At the beginning of the United States program, environmentalists and fear mongers rallied to stop the development of nuclear power and the construction of nuclear reactors within the borders of the Unite States.

The first privately built commercial reactor in the United States was Dresden Station, reactor number 1, Grundy County, Illinois, that opened in 1959. Twenty years later in March 1979 a partial meltdown of the core occurred in the Three Mile Island reactor number 2, near Harrisburg, Pennsylvania. Because of this accident the construction of new nuclear power plants in the United States began shutting down amid a crescendo of irrational fear, ignorance, and propaganda. We are lucky that the United States had built 104 nuclear reactors that supply about 20 percent of our electrical needs before the shutdowns started. In 2007 President Bush said he wanted to start building nuclear reactors in the United States again as soon as possible.

NUCLEAR POWER FOR EVERYONE EXCEPT U.S.

Fortunately, the stoppage was a United States problem because other countries continue to build nuclear power plants. For example, Lithuania gets 80 percent of its electricity from nuclear power plants; France gets 77 percent. Lithuania and France are at the top of the list, percentage wise, of the 31 countries that operate nuclear power plants. Only in the United States does there seem to be a fear of nuclear power.

As an example, consider the disastrous meltdown of one of the four Chernobyl reactors that caused the death of dozens and perhaps hundreds and spread radioactive fallout as far away as Scandinavia. The other three remaining reactors are scheduled to be shut down; however, the Ukrainians are going to build two new nuclear reactors to make up for their loss of electricity.

ELECTION OF 2004

The shutdown of nuclear power plants is a political issue with the anti-nuke crowd, led primarily by Democrats, attacking any effort to get new plants on line. The differences between the Republicans and Democrats vis-à-vis nuclear power production was made clear by the positions put forward during the 2004 presidential campaign. The Republican campaign called for building new nuclear power plants and storing the radioactive by-products in Yucca Mountain, Nevada. The Democrats would not allow the construction of new nuclear power plants and would close down Yucca Mountain as a nuclear waste storage facility.

OBAMA CHANGES COURSE

President Barrack S. Obama signaled that he wanted nuclear power plant construction to start up again in the United States. This is a fortunate and welcome reversal of the long-held Democrat position.

A nuclear power plant is cheaper to operate than a fossil fuel power plant. It is cleaner, there being absolutely no green house gases released to the atmosphere. The fallout from the stacks of power plants in the Midwest that burn coal creates a problem in New England where it is said to fall to earth as acid rain. Replacing those coal-fired furnaces of the Mid-West with nuclear fission furnaces would solve the acid rain problem and relieve the contentious bickering between the Midwest and New England over acid rain.

A WORLD WIDE EFFORT NEEDED

The United States and all other countries need to push ahead to build nuclear power plants in such number that everyone on earth will have access to electricity. New and advanced designs have given us reactors of high efficiency that are meltdown proof. The safe storage of radioactive nuclear reactor byproducts is not a problem if men of good will come together to solve it.

MOTHER EARTH

Our cool blue planet, earth, which has passed through several known ice ages, is in reality, a "dark star." A dark star is a galactic entity about the size of our earth that contains enough mass so that the inward pull of gravity is strong enough to melt the interior, but not strong enough to cause it to glow (fuse) like the sun. Indeed, the interior of the earth is mostly molten iron, called magma, that at the center of the earth, according to best estimates, reaches a temperature higher than that at the surface of the sun.

An ideal set of circumstances allows living organisms, such as humans, to live comfortably on the surface of the earth in a nourishing environment. Several phenomena account for this. In the first place, the relatively large size of the earth creates the gravitational pull that keeps the atmosphere from drifting off into space. Secondly, water covers about 71 percent of the earth's surface. The oceans, too, would drift into space if it were not for the gravitational pull that keeps them fixed to the surface of the earth. If the earth lost its air and water, it would be a "dead" planet such as our moon. Such an environment would be so hostile to life that we all would die.

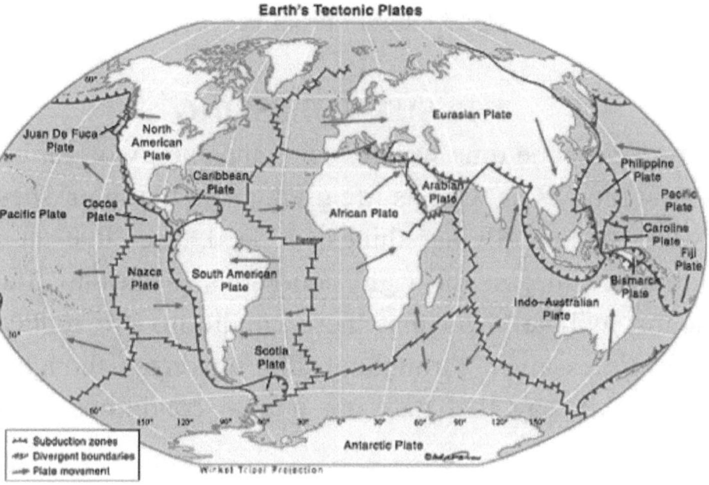

Figure 4.5. The Earth's Tectonic Plates. Source: Maps com

A third factor helps to cool the high temperature of our dark star. That is the crust of the earth. The crust is made up of seven large continental tectonic plates and several smaller plates that float on top of the magma. The plates move about very slowly relative to each other guided by currents created deep within the molten core of the earth. New crust is created along "rift zones" where the plates split apart as lava wells up between them from inside the earth. Rift zones occur in Iceland and at the bottom of the Atlantic Ocean, among others. New crust is also created when volcanoes spew lava through holes in the crust. For example, Kilauea Volcano on the island of Hawaii in the Hawaiian Islands has been erupting since 1983 and has added hundreds of acres of new land to the island. The crust transmits energy (heat) from the center of the earth to the oceans and atmosphere that transmit it into space. This loss of heat goes on all the time at about the same rate in what is called a "steady state" process.

THE SUN'S BLESSING

At the same time we are losing the energy of our earth, the sun blesses us with its life giving rays that nourishes all of the earth's living things. Indeed, we live at the confluence of just the right conditions to sustain life. A scientist will tell you that these conditions are replicated many times over throughout the universe, though the faithful will say that they are unique conditions brought about by God.

GEOTHEMAL ENERGY

Beneath the crust of the earth is virtually an inexhaustible supply of energy. There is enough energy (heat) to supply the needs of all mankind for billions of years. Energy derived from the core of the earth is called "Geothermal Energy." The word geothermal comes from the Greek word 'geo' meaning earth and 'therme' meaning heat.

The center of the earth is approximately 4,000 miles below its surface. The temperature at the center of the earth is said to be about 9,000° F.

The crust of the earth is 3 to 35 miles thick. The deeper thickness of the crust is located on the surface where mountains have "piled up" on the continental land masses, whereas, the shallower thickness is found where the crust is splitting apart along rift zones.

RING OF FIRE

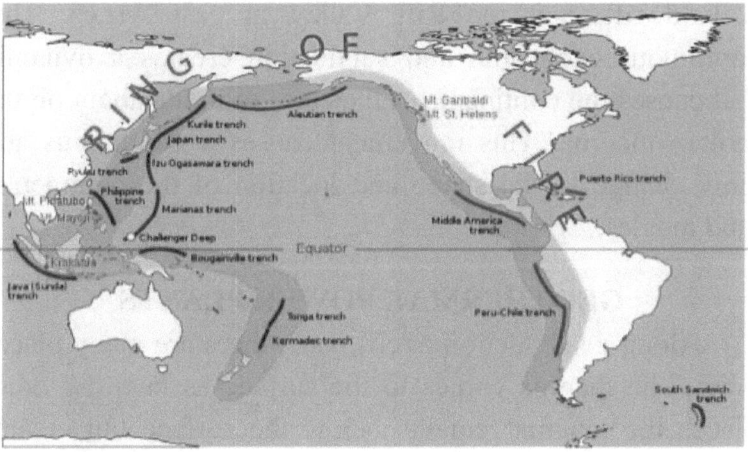

Figure 4.6. The Ring of Fire. Source: WorldAtlas.com

The best opportunity to harness geothermal energy occurs along the edges of the tectonic plates as they collide. This collision pushes magma to the surface and causes the crust of the earth to shift (the cause of earthquakes) at the rate of 1 to 2 inches a year (about the rate of growth of your fingernails). The most spectacular display of this process takes place around the "Ring of Fire," which is caused by the Pacific Ocean plate, as it expands, melting as it slides under (subduction) the continental plates. As a plate slides under a continental plate it creates a deep trench where it goes under. These trenches are the deepest places in the oceans and are found along the edges of the continents. Subduction is caused by spreading, or rifts, that take place within a plate. The rift expands the surface area of the plate, thereby, pushing it under an adjacent continental land mass.

Examples of rifts are: the Atlantic Ocean rift zone, the Iceland rift, and the Rift Valley in east Africa. This continuous expansion and subduction creates a dynamic that causes the continents of the earth to move about on the molten magma. This movement causes a continuous and slow change in the shape and location of the continental land masses.

GEOTHERMAL POWER PLANTS

Along a subduction or rift zone, there are many places where the magma comes to the surface as lava. At other places the magma remains below the surface but creates "hot spots" that spew out steam or hot water or both steam and hot water. Such places are most favorable for the production of electricity. In fact, geothermal wells are now producing over 8,000 megawatts of electricity and there is a potential for much more to be developed, especially in the western part of the United States and Hawaii.

There are two kinds of geothermal formations that are being developed for the production of electricity.

DRY STEAM GEOTHERMAL PLANT

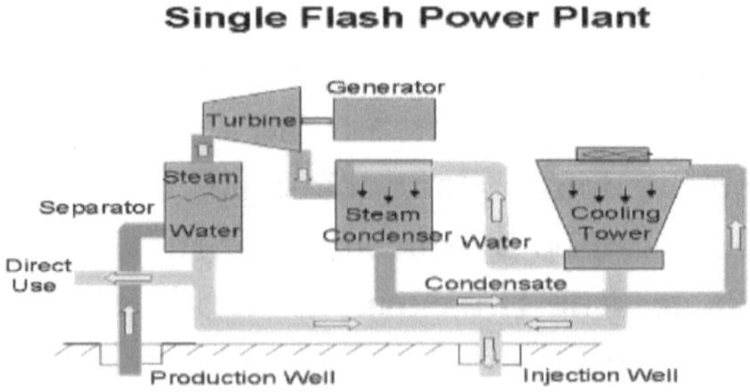

Figure 4.7. Schematic of a Flash Power Plant (Dry

Steam)

The first type of geothermal generator uses dry steam from the geothermal well to turn an electrical turbine. "The Geysers" located about ninety miles north of San Francisco is the largest deposit of dry stream in the world. Geothermal power plants located at "The Geysers" have been on line since the 1960s. Italy is the only other place where dry stream exists in an amount useful for exploitation. The dry steam is pumped directly into a turbine that turns a generator to produce electricity.

BINARY CYCLE GEOTHERMAL POWER PLANT

The second type of geothermal power plant, Binary Cycle Power Plant, uses hot water from a well that is not hot enough to produce dry steam. The water is transferred into a vacuum chamber where it "flashes" into steam that is fed into a turbine to produce electricity.

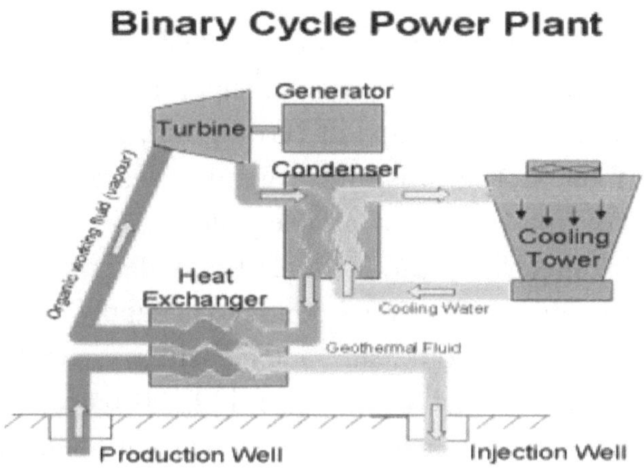

Figure 4.8. Schematic of a Binary Cycle Power Plant

The Binary Cycle Power Plant uses lower temperature water from a well to produce steam. This is done by using a liquid with a low boiling point such as isopentane, which "flashes" to produce steam that drives the turbine of an electric generator. (Schematics are from the Department of Energy.)

A geothermal power plant is clean, reliable, and flexible in that new generating units can be added on as more electricity is needed.

THE PUNA, HAWAII GEOTHERMAL VENTURES

It would seem to have been a simple process for the Hawaiian community to come together to solve its need for additional electrical power. It would have been easy to harness the free geothermal power available for the taking at Kilauea Volcano. Instead of getting the much needed source of electricity in a few years, it took decades.

The Puna Geothermal Ventures (PGV) was started in the 1970s because of the expense of supplying electricity to the sparsely populated Island of Hawaii during the Arab oil boycott of 1973. The prospect of getting steam out of the Kilauea Volcano proved to be adequate for driving steam turbines for electricity production. It was determined that Kilauea was able to provide a continuous supply of cheap, clean electricity from an inexhaustible supply of geothermal energy that would save 429,600 barrels of oil each year. Plans were made to expand capacity and supply the islands of Maui and Oahu through an undersea cable.

THE OBSTRUCTIONISTS

During the twenty years between 1973 and 1993 when the first 30-megawatt power plant went on line for commercial electricity production the environmental obstructionists were in full battle array in their efforts to shut it down. There was litigation and hundreds of arrests for civil disobedience. Prestigious academicians wrote thousands of articles and papers decrying the PGV as the embodiment of willful destruction and evil.

In June of 1991 there was an accidental release of hydrogen sulfide into the atmosphere. The Sierra Legal Defense Fund (SLDF), in coalition with twelve environmental, cultural, neighborhood, and business groups responded by filing a lawsuit against the State of Hawaii. The SLDF lawsuit said "…this is the decisive stroke in our campaign to protect the Wao Kele O Puna rainforest."

The Hawaii Health Department found that no risk to the population, the rain forest, or to geothermal development existed from the June 1991 release of hydrogen sulfide into the atmosphere and allowed the drillers to go back to work.

In April 2004 Ormat Nevada, Inc., an Israeli Company, bought PGV from Constellation Energy of Baltimore for $71 million and vows to expand production by adding more generators. It is estimated that PGV can eventually produce 500 to 600 megawatts of electricity. That is enough to supply the electrical needs of the entire State of Hawaii.

Figure 4.9. Kilauea Geothermal Separator
Source: Hawaii247.org

In December 2008, Lieutenant Governor Duke Aiona and Lucien Y. Breniki, Chief Technical Officer of Ormat Corporation, met at Puna to celebrate fifteen years of successful operation of the geothermal electricity generator.

Lieutenant Governor Duke Aiona declared, “The era of oil is over.”

A plant technician said, “Our plant is closed loop with five production wells drawing hot steam from beneath Kilauea Volcano’s east rift zone and three injection wells pumping the steam back into the earth.”

Mr. Breniki renewed the confirmation that there is enough geothermal energy available in Kilauea Volcano to supply the entire state of Hawaii with electricity.

The successful operation of the PGV has not quieted the outcry against it which still rages on.

DIGGING DEEPER FOR GEOTHERMAL ENERGY

There is another source of geothermal energy.

If we could reach down deep enough to tap the heat of the molten core of the earth, all the energy needs of mankind would be solved forever. If we could harness the energy of the core of the earth, the only expense to making all the electricity needed would be the cost of maintenance and distribution: That is, of course, after drilling and construction are done.

This untapped source of geothermal energy is available three to six miles below the surface of the earth. It is the hot dry rock substrata which has no water that can be pumped up to drive a turbine. It can be exploited by drilling down to the dry rock and pumping in water and taking out steam in an uptake pipe to produce electricity.

New drilling methods make this source of geothermal energy a real possibility. New laser drills allow the sinking of a well much less expensive and as much as a hundred times faster than mechanical drills.

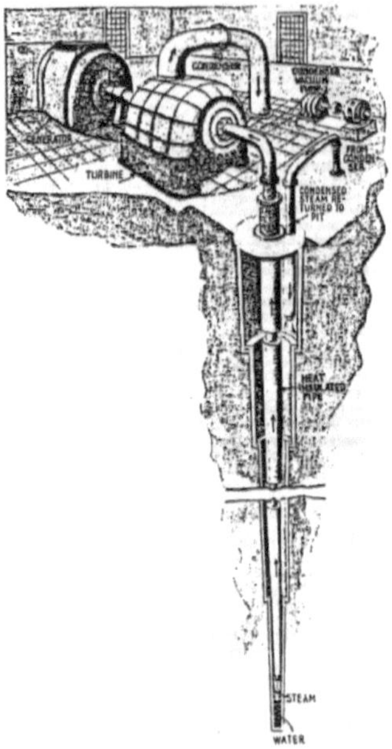

Figure 4.10. Nicola Tesla's 1931 concept
of a dry rock power plant

Tesla outlined the arrangement for one of the great terrestrial-heat power plants of the future. Water is circulated to the bottom of the shaft, returning as steam to drive the turbine, and then returned to liquid form in the condenser, in an unending cycle. "The internal heat of the earth is great and practically inexhaustible,'' Tesla wrote.

The idea of connecting our energy needs to the core of the earth is not new. Futurists such as Tesla had the dream and knew that we could have an inexhaustible source of energy if there was the will and ability to dig deeper.

By harnessing the nuclear and geothermal energy available at our fingertips, there is no reason that we cannot have an inexhaustible supply of electrical energy.

ADEQUATE SUPPLY OF FOSSIL FUEL

Despite the alleged shortage that pushed up the price for a barrel of crude oil to record highs, Abdullah S. Jum'ah, Chief Executive Officer of the Saudi Arabian Oil Company said on September 13, 2006, that only 18 percent of the world's oil reserves have been tapped and that there is enough crude to power the world at current levels of use for the next 140 years.

No doubt Abdullah is right, but the big squabble today is not about using crude to power the world at current levels. It is about reducing consumption to reduce green house gases. Environmentalists demand an alternative means of power production that does not depend upon burning oil.

A SILICON VALLEY SOLUTION

Tesla Motors, Incorporated. opened a Silicon Valley startup company that has produced a high performance electrical car that runs on more than 6,000 lithium-ion batteries. The company, named for inventor Nicola Tesla, has created a sports car with a top speed of 135 miles per hour (MPH) with acceleration from 0 to 60 MPH in 4 seconds and a range of 250 miles (430 kilometers) on each charge of the batteries. The first cars hit the road in the summer of 2008. The car was made in England by the Lotus Automobile Company.

Tesla Motors is working on a sedan called "White Star" that the company introduced last year. The car will cost 1 to 2 cents per mile to operate on the open road. The lithium-ion battery pack will last for 100,000 miles and is recharged by plugging it into an electrical outlet. The White Star is expected to roll off the assembly lines in New Mexico and California late 2009 and will cost about $60,000 each.

Figure 4.11. Tesla Motors White Star sedan
Source: Auto blog

PUTTING THE OIL PRODUCERS OUT OF BUSINESS

The implication of this development, as it comes to fruition within the next few years, is that it signals the death knell of the world's total dependence on crude oil for cars. The Organization of Petroleum Exporting Countries (OPEC) and other nations that produce large quantities of oil for sale in today's markets will lose power, prestige, and financial primacy.

The advent of the Tesla Motors Electric Car will be another free gift to the world from the minds of American and Silicon Valley geniuses. Just plug the car into your house current over night and be ready to hit your commute to town or the office in the morning.

We have at our fingertips the means, the tools, and

the science that provides us with the power to shape our lives and environment to make it a safer, cleaner, and better world for ourselves and our children.

CHAPTER 5: WAR

On September 11, 2001, radical Muslims intensified a jihad or "holy war" against western civilization, the United States, and Jews. They kicked off the new phase of their war of terror by hijacking four United States jetliners loaded with innocent passengers and crashing them into the two World Trade Center towers and the Pentagon. A fourth plane bound for The White House or Capitol Building in Washington, District of Columbia, crashed into a field in Pennsylvania. Three thousand innocent men, women, and children died.

This unprovoked attack was planned, financed, and executed under the direction of a wealthy Saudi Arabian, Osama bin Laden. Osama struck the United States from his headquarters in Afghanistan. Through the use of his al-Qaeda terrorist network, bin Laden gained control of Afghanistan by merging with a radical and repressive Muslim sect called the Taliban that gained control of the Afghan government by force of arms in 1996.

LIBERATION OF AFGHANISTAN

After the unprovoked assault by al-Qaeda of September 11, 2001, upon the United States, President George W. Bush declared, and the Congress approved, war on al-Qaeda and its network of terrorists. Within weeks the Taliban was defeated and the Afghanistan people were liberated from the forces of oppression. The al-Qaeda was decimated and Osama bin Laden, fled into the caves of northeastern Afghanistan where he was holed up in Afghanistan's tribal region along the Pakistani border. (He was eventually tracked down and killed by U.S. Navy Seals on May 2, 2011.)

HAMID KARZAI

The people of Afghanistan chose Hamid Karzai, a charismatic leader, to turn the nation away from Taliban extremism. Karzai unified the nation and legitimized its government by holding national elections for president and Delegates to its National Assembly. He set up Afghani Armed Forces on a national level and continues to hunt down and kill al-Qaeda and Taliban extremists. He suppressed the power of the tribal war lords and invited them to come into the national government.

Karzai cooperated closely with the North Atlantic Treaty Organization (NATO) forces that helped him hunt down the Taliban.

Karzai had expressed a willingness to get rid of the poppy fields that supply 90 percent of the world's heroin, and replace them with a different cash crop. This remains a daunting task because the poppy plant, the source of heroin, is Afghanistan's largest cash crop.

The Afghan government distributed saffron bulbs to farmers so they can grow saffron instead of poppies. Farmers in the region along the Iranian border say the new crop is better suited to their religion and even more profitable. Saffron production increased every year in the three years prior to this writing. The spice is a key ingredient in the cuisine of the region, adding a distinctive yellow color to rice and other dishes. [13]

CONQUEST OF IRAQ

In March 2003, President Bush of the United States and Prime Minister Tony Blair of the United Kingdom conquered Iraq to enforce United Nations resolutions and to rid the world of an evil dictator, Saddam Hussein (hanged by the Iraqi Court on December 30, 2006) who had the blood of hundreds of thousands of Iraqis on his hands.

The Iraqi people voted overwhelmingly to form a new government and wrote a constitution based on the United States Constitution. The government was based upon a coalition between the Shiites, Kurds, and Sunni religious factions. The Iraqi constitution provided that all factions would be treated fairly and that civil law, rather than religious law, would rule.

BUSH MISTAKES

The conquest of Iraq was completed in a matter of days and was one of the most successful military campaigns in history. Unfortunately, the president did not heed the advice of his Army Chief of Staff, General Eric K. Shinseki, that 300,000 boots on the ground would be needed to pacify Iraq. Instead, the president went along with Donald H. Rumsfeld, his Secretary of Defense, who fired General Shinseki. As a result of this mistake, several years of bungling and inefficiency took over United States policy and operations in Iraq.

AL-QAEDA COUNTER ATTACK

At the same time, al-Qaeda from all over the world began to infiltrate Iraq with the aim of defeating United States and Coalition Forces before a legitimate Iraqi government could be formed. While still in its infancy, in the midst of mortar attacks and suicide bombers, the government of Iraq held steadfast to its goal by voting at each step along the way to achieve a democratic government. At the same time, Iraqi battalions were trained by United States and Coalition Forces to assume the role of defending the new government of Iraq and defeating the jihad, but could not get the job done.

ELECTIONS OF 2006

The war dragged on from 2003 until the mid-term elections of 2006.

In a bitter defeat for President Bush, the Democrats took control of both the Senate and House of Representatives. The day after the election Rumsfeld resigned his post as Secretary of Defense. The Democrats launched a campaign to pull United States Armed Forces out of Iraq. President Bush refused to withdraw from Iraq. Instead, he demanded and got from the Congress, an increase (surge) of 30,000 troops in a last ditch effort to conquer al-Qaeda and turn defense of the country over to the Iraqi Army. The president bought time for his surge at the price of a progress report from the officer in charge, General David H. Petraeus, on September 15, 2007.

The surge succeeded.

Because of the success of the surge, the al Qaeda was decimated, the Iraqi Army was strengthened, détente reached between the Shiite majority and Sunni minority. New elections were held in 2009.

OBAMA POLICY

The stated policy of President Barrack Obama was to withdraw United States Armed Forces from Iraq. An end U. S. Army occupation was scheduled for 2011.

Instead, the war in Afghanistan was ramped up. Several more brigades of troops were scheduled to be posted to Afghanistan.

ENDNOTES

1. "Greenspan admits 'mistake' that helped crisis," *Associated Press*. October 23, 2008.

2. *Encyclopedia Britannica*, 1960 edition, vol.19, page 725.

3. *Population Bulletin* Vol. 63, No. 3 September 2008.

4. *Encyclopedia Britannica*, 1960 edition, vol. 14, pages 744 & 745.

5. Ibid., vol. 4, page 396.

6. *New World Dictionary of American English*, 3rd College Edition, page 369.

7. *Population Bulletin,* Vol. 63, No. 3 September 2008.

8. "Ukrainian Famine," ibiblio.org.

9. "Great Chinese Famine," Wikipedia.org

10. Keim, Brandon. "Bush Triple Funds for AIDS, Malaria and Tuberculosis," Wired.com, July 31, 2008.

11. Toth, Mike. "Stanford's Nuttiest Tenured Turkey," *The Stanford Review*, March 10, 1998.

12. *Encyclopedia Britannica*, 1960 edition, volume 19, page 725.

13. *The Washington Times*, January 26, 2009, page 23.

INDEX

L

M

N

O

P

ABOUT THE AUTHOR

Daniel James Myers, Jr. (12/20/28 – 11/23/11)

Daniel James Myers, Jr., graduated from the United States Military Academy, West Point,* New York, with a bachelor of science degree in Engineering and from Tulane University, New Orleans, Louisiana, with a master of science degree in Nuclear Physics.

After graduation, he was stationed in New Mexico for five years with the Defense Atomic Support Agency where he did research, development, and testing of nuclear warheads during the 1960s.

He was a decorated combat infantryman who fought in both the Korean War and Vietnam Wars. He served in paratrooper units in Japan and the United States. He was a graduate of the United States Army Command and General Staff College, Fort Leavenworth, Kansas.

Retiring after twenty years in the Army as a lieutenant colonel, Mr. Myers opened "Dan Myers Real Estate" in Knoxville, Tennessee. It was a very successful operation until interest rates rose into the high teens during the late 1970s. He sold the company and moved to Hawaii in 1978.

In Honolulu, Myers went to work for Merrill Lynch as a financial consultant. The 1980s, the Reagan years, ushered in a period of low taxes, economic growth, and expansion. It was an exciting and profitable time to be invested in stocks as the market moved to all-time highs. He retired from Merrill Lynch in 1991 as vice president.

Daniel lived with his wife, Beth, in Hawaii until his death in 2011. They have five children and nine grandchildren. Daniel has a daughter, son and an adopted son from his first marriage and inherited a daughter and son by his second marriage to Beth.

West Point classmate, Jack Craigie, stated: "Dan was more than a hero in 1951. He was a workhorse. *Code*Breakers* was a good movie, but it left out Dan's pivotal role. For almost four months Danny was a leading investigator of a large cheating enterprise. In his role on the Honors Committee, Dan did a job that has never before existed with perseverance, judgment, and maturity beyond his years. He did so under the most stressful and bewildering circumstances. Indeed, to say Danny and West Point were covering new territory would be an understatement of epic proportions."

*The cheating scandal resulted in the expulsion of 83 cadets from the academy by President Truman.

All net proceeds from the sale of this book
will be donated to the
Army Emergency Relief Fund
in memory of Daniel Myers

Made in United States
Troutdale, OR
11/03/2026

56844695R00076